Biomedical NLP Workshop 2017

Varna, Bulgaria
8 September 2017

ISBN: 978-1-5108-5546-5

Proceedings of the
Biomedical NLP Workshop

associated with
The 11th International Conference on
Recent Advances in Natural Language Processing
(RANLP 2017)

8 September, 2017
Varna, Bulgaria

BIOMEDICAL NATURAL LANGUAGE PROCESSING WORKSHOP
ASSOCIATED WITH THE INTERNATIONAL CONFERENCE
RECENT ADVANCES IN
NATURAL LANGUAGE PROCESSING'2017

PROCEEDINGS

Varna, Bulgaria
8 September 2017

Designed and Printed by INCOMA Ltd.
Shoumen, BULGARIA

Preface

Biomedical NLP deals with the processing of healthcare-related text—clinical documents created by physicians and other healthcare providers at the point of care, scientific publications in the areas of biology and medicine, and consumer healthcare text such as social media blogs. Recent years have seen dramatic changes in the types and amount of data available to researchers in this field. Where most research on publications in the past has dealt with the abstracts of journal articles, we now have access to the full texts of journal articles via PubMedCentral. Where research on clinical documents has been hampered by a lack of availability of data, we now have access to large bodies of data through the auspices of the Cincinnati Children's Hospital NLP Challenge, the i2b2 shared tasks (www.i2b2.org), the TREC Electronic Medical Records track, Clinical TempEval series of tasks, the US-funded Strategic Health Advanced Research Projects Area 4 (www.sharpn.org) and the Shared Annotated Resources (ShARe; https://sites.google.com/site/shareclefehealth/taskdescription; www.clinicalnlpannotations.org) project. Meanwhile, the number of abstracts in PubMed continues to grow exponentially. Text in the form of blogs created by patients discussing various healthcare topics has emerged as another data source, with a new perspective on healthrelated issues. Connecting the information from the three main sources in multiple languages to the scientific community, the healthcare provider, and the healthcare consumer presents new challenges.

The Biomedical Natural Language Processing at RANLP 2017 provided a venue for presentations of current work in this field. The topics of papers presented at the workshop included information retrieval, part-of-speech tagging, multi-part knowledge frames population, extraction of numerical described values, resource creation, entity-centric information access, named entity recognition, confidence estimation for protein-protein relation discovery and association rule mining of clinical text and the biomedical literature.

The Workshop Organizers

The BioNLP Workshop associated with RANLP'17 is organised by:

Svetla Boytcheva (IICT, Bulgarian Academy of Sciences)

Kevin Bretonnel Cohen (University of Colorado School of Medicine)

Guergana Savova (Harvard Medical School and Boston Children's Hospital)

Galia Angelova (IICT, Bulgarian Academy of Sciences)

The event is partially supported by:

National Scientific Fund, Ministry of Education and Science, Bulgaria

Program Committee:

Galia Angelova (Bulgarian Academy of Sciences)
Svetla Boytcheva (Bulgarian Academy of Sciences)
Kevin Cohen (U. Colorado School of Medicine)
Noa P. Cruz Diaz (Group of Research and Innovation in Biomedical Informatics, Biomedical Engineering and Health Economy. Institute of Biomedicine of Seville/ Virgen del Rocío University Hospital / CSIC / University of Seville)
George Giannakopoulos (NCSR Demokritos & SciFY NPC)
Agnieszka Mykowiecka (Polish Academy of Sciences)
Preslav Nakov (Qatar Computing Research Institute, HBKU)
Ivelina Nikolova (Bulgarian Academy of Sciences)
Georgios Petasis (NCSR "Demokritos")
Guergana Savova (Harvard Medical School and Boston Children's Hospital)
Frédérique Segond (Viseo Research)
Dimitar Tcharaktchiev (Medical University - Sofia)

Reviewers:

Ekaterina L. Chernyak (Higher School of Economics, Moscow)
Dmitry Ilvovsky (Higher School of Economics, Moscow)
Natalia Korepanova (Higher School of Economics, Moscow)

Table of Contents

Document retrieval and question answering in medical documents.
A large-scale corpus challenge.

Eric Curea

Research Institute for Artificial Intelligence
"MIHAI DRAGANESCU",
Romanian Academy
`eric@racai.ro`

Abstract

Whenever employed on large datasets, information retrieval works by isolating a subset of documents from the larger dataset and then proceeding with low-level processing of the text. This is usually carried out by means of adding index-terms to each document in the collection. In this paper we deal with automatic document classification and index-term detection applied on large-scale medical corpora. In our methodology we employ a linear classifier and we test our results on the BioASQ training corpora, which is a collection of 12 million MeSH-indexed medical abstracts. We cover both term-indexing, result retrieval and result ranking based on distributed word representations.

1 Introduction

Automatic key-wording is the process of enriching text documents with pre-specified classes (topics or themes). The primary motivation is that in information retrieval one can easily use these keywords for automatically filtering and obtaining a subset of documents form a large-scale corpus, documents that share common traits linked to their domain, topic, title, publication source, authors, etc. As such, automatic key-wording and document indexing (based on these keywords) helps people to find information in huge resources.

Currently, most of the on-line information is available in the form of unstructured documents and this is unlikely to change in the foreseeable future. Though, several initiatives to force users into manually labeling their on-line publications using specialized markup have been proposed (one good example is Google Markup Language[1]), scientific publications are unlikely to be subject to such annotations, mainly because they employ printable formats such as Postscript and PDF (which, in fortunate situations, can be converted into plain text).

Thus, NLP task such as unsupervised document clustering represents a key-task in information retrieval. Due to the increased availability of documents in digital form and the ensuing need to access them in flexible ways, content-based document management tasks (collectively known as information retrieval IR) have gained a prominent status in the research community in the past decade. The task of Document classification or document categorization, the activity of labeling natural language texts with thematic categories from a predefined set, is very important and still evolving thanks to increased applicative interest and to the availability of more powerful hardware.

To accomplish the task of document classification, an increasing number of computational and statistical approaches have been developed over the years, to mention a few: Suport Vector Machines (SVMs) (Manevitz and Yousef, 2001; Joachims, 1998), maximum entropy (Ratnaparkhi, 1998; El-Halees, 2015), word-distributional clustering (Baker and McCallum, 1998), weighted K-Nearest-Neighbor classification (Han et al., 2001; Larsen and Aone, 1999), linear classifiers (Lewis et al., 1996), Naive Bayes methods (McCallum and Nigam, 1998), artificial neural networks (Zhang and Zhou, 2006; Collobert and Weston, 2008; Lai et al., 2015), decision trees (Lewis and Ringuette, 1994).

Our work is focused on automatic labeling of medical text, using Medical Subject Headings (MeSH)[2] terms (Rogers, 1963) (see section 3)

---

[1] https://developers.google.com/search/docs/guides/intro-structured-data - accessed 2017-05-18

[2] https://www.ncbi.nlm.nih.gov/pmc/articles/PMC35238/

Proceedings of the Biomedical NLP Workshop associated with RANLP 2017, pages 1–7,
Varna, Bulgaria, 8 September 2017.

and information retrieval for question answering based on the analysis of article abstracts (see section 4). The training, evaluation and test datasets used in the validation of our procedure are part of the BioASQ[3] (Tsatsaronis et al., 2012) evaluation campaign.

2 Corpus description

The train set corpus contained articles from the free on-line repository PubMed [4]

The training data is composed of a very large number of documents collected from PubMed, which have been semi-automatically annotated with MeSH terms, with the help of human curators. Aside from the MeSH terms, each entry in the dataset contained important meta-data such as: the title of the paper, the journal where the paper was published, publishing year and the paper's abstract.

The training set is JSON-encoded and contains the following fields for each article:

1. pmid : An unique identifier assigned to each paper - used for internal evaluation purposes;

2. title : The original title of the article

3. abstractText : the abstract of the article,

4. year : the year the article was published,

5. journal : the journal the article was published, and

6. meshMajor : a list with the major MeSH headings of the article.

For clarity, we also provide an excerpt from the training data, presenting the structure af each article collected in the large-scale corpus:

```
{"articles": [{"journal":"journal..","
    abstractText":"text..", "meshMajor
    ":["mesh1",...,"meshN"], "pmid":"
    PMID", "title":"title..", "year":"
    YYYY"},..., {..}]}
```

To offer a better view over the training data, we must specify that the total number of articles is 12,834,585, published in over 9,000 journals, with an average of 1,421.64 articles in each journal, published from 1946 to 2016 with most articles (over 600,000) selected from 2014, a distribution

Table 1: Label distribution over training data

labeled documents	distribution	percentage
$>$1,000,000	10	0.04%
500,000 1,000,000	7	0.03%
100,000 500,000	137	0.49%
50,000 100,000	223	0.80%
10,000 50,000	2,240	8.07%
1,000 10,000	10,053	36.20%
$<$1,000	15,103	54.38%
total	27,773.00	

of 12.66 average MeSHes per article, going from the MeSH "humans" with an occurrence of over 8 millions to MeSHes like "tropaeolaceae" that only occur once, yielding a MeSH coverage of 27,773 MeSHes composed either from a single word of a construct like "magnetic_resonance_imaging". All this in a total of 20.5GB (plain/text) and 6.29GB (compressed data). Table 2 provides generic information regarding frequent versus uncommon MeSHes, while table 1 captures the "spread" of the MeSHes throughout the training data.

As can easily be seen from table 1, 10 of the frequent MeSHes like "humans", "male", "female" or "animals", are used to label more the 1M documents, only 7 fall within the 500K-1M range and 360 between 50K and 500K (we further refer to them as category A). On the opposite side, 2K MeSHes are found in 10K-50K documents, 10K MeSHes in 1K-10K documents and more than 15K MeSHes have an occurrence of less than 1K (category B). The high occurring MeSHes (category A) represent less than 2% of the total number of labels, which indicates that in most cases any ML system will most likely not be able to model the rest of 98% of the labels based on this corpus. To clarify our previous statement, it is expected that most classifiers will have a small recall for 98% of the labels, mainly because the objective of minimizing the "overall" accuracy is easily achieved by preferring not to emit any label rather than incorrectly classifying documents with bad labels and only for less than 2% of the total number of labels the systems will have a chance of a high recall.

3 Automatic MeSH labeling

Currently, there are 28,489 descriptors in MeSH 2017 that were used in the creation of the training data. However, due to the unbalanced occurrence

- accessed 2017-05-18

[3]http://bioasq.org - last accessed 2017-05-09

[4]https://www.ncbi.nlm.nih.gov/pubmed/ - accessed 2017-04-29

Table 2: MeSH distribution

ID	MESH	count	ID	MESH	count
1	humans	8,103,280	27	kinetics	366,997
2	male	5,351,269	28	cell_line	331,436
3	female	5,169,536	29	surveys_and_questionnaires	316,552
4	animals	3,932,184	30	rna+messenger	314,638
5	adult	3,119,705	31	dose-response_relationship+drug	313,386
6	middle_aged	2,782,688	32	reproducibility_of_results	285,023
7	aged	1,936,405	33	infant+newborn	283,249
8	adolescent	1,219,944	34	mutation	278,419
9	rats	1,116,126	35	united_states	272,593
10	mice	1,045,215	36	brain	269,598
11	child	826,020	37	rats+sprague-dawley	265,472
12	time_factors	793,584	38	sensitivity_and_specificity	264,091
13	aged+80_and_over	636,261	39	prognosis	259,335
14	molecular_sequence_data	590,276	40	in_vitro_techniques	258,033
15	treatment_outcome	571,489	41	age_factors	254,441
16	retrospective_studies	547,781	42	liver	248,866
17	child+preschool	510,539	....		....
18	young_adult	494,101		ephemerovirus	5
19	risk_factors	450,495	....		....
20	follow-up_studies	447,572		zigadenus	4
21	cells+cultured	428,059	....		....
22	amino_acid_sequence	395,146		cytophagaceae_infections	3
23	prospective_studies	394,813	....		....
24	pregnancy	392,281		duboisia	2
25	infant	387,000	....		....
26	base_sequence	385,031		childhood-onset_fluency_disorder	1

of terms combined with the large scale of the corpus, we ran our experiments on a smaller sub-set of MeSH terms, composed of only 154 most frequent items.

In the classification process we took into account as much information as we can and have access to, about each document in the large-scale corpus. The title of a document usually holds key information about the content of the document. The journal in which it was published is likely to carry weight in the label assigning process as only specific types of documents can be published in certain types of journals. The year in which the document was published will tell the system if the information retrieved from the document has a chance of not being up to date or it might be completely outdated and superseded by more recent research, in which case the system should at least try to see if newer publications might hold better results or more important supplementary information. The abstract text is the place where the system can spend most processing time and apply as many tests, approximations and refinements, because this is the place where most articles condense the biggest amount of relevant information about the content of the document. Of course finding possible relevant information in the abstract text is only part of the equation. The more important part is determining relevant relations between different relevant lexical tokens, the location of the information segments, distance between the different relevant lexical tokens inside the abstract, number of occurrences, similarity to the information determined in the question (W2V, cosine similarity(Steinbach et al., 2000)).

All the input features were treated in a bag-of-words manner, from which we removed any feature (word) with an occurrence rate lower than 100. This threshold of 100 was selected after testing different limits that yielded either too few features left to test with or too low occurrence rate for the feature to be relevant. Initially, our training data contained 7,466,119 unique features and the pruning process reduced this number to only 123,255. For the classification task we employed an ensemble of linear classifiers. Each possible output MeSH was associated with a classifier, which was trained in a 1-vs-all style to predict if the system should or should not assign that label, based on the input features. The output of the linear model ranged from -1 (do not assign a label) to 1 (assign a label) and was computed using Equation 1, with w computed using the delta-rule (Equation 2):

$$y = \sum_{1}^{n} w_n \cdot x_n \qquad (1)$$

$$\Delta w_k = \alpha \cdot (t - y) \cdot x_k \qquad (2)$$

where
y is the output of the classifier
t is the desired output of the classifier (-1 or 1)
x_i is the ith input feature
w_i is the weight of the i-th input feature
α is the learning-rate (set to 10^{-3})

When we trained our ensemble of classifiers we divided our training data into 9/10 for training and 1/10 for development, while trying to preserve as best as possible the initial distribution for each of the labels in both sets. Training was done iteratively (compute new value for w using the training set and measure accuracy on the development set) and the stopping condition was not to have any improvements on the development set for more than 20 iterations. At the end of the training process we kept the w that achieved the highest accuracy on the development set.

Table 3: Labeling results

System	MiP	MiR	Acc.
Sequencer	0.0920	0.0964	0.0494
Default MTI	0.6148	0.6286	0.4594
Our System	**0.7681**	**0.1472**	**0.1381**
DeepMeSH4	0.6671	0.6289	0.4839
MZ1	0.6495	0.3985	0.3299
DeepMeSH3	0.6898	0.6170	0.4877
DeepMeSH2	0.6895	0.6432	0.5059
DeepMeSH1	0.7025	0.6282	0.5025
DeepMeSH5	0.7198	0.6122	0.5024

Table 3 shows the accuracy (Acc), Micro Precision (MiP) and Micro Recall (MiR) of our system, measured on one of the datasets. It also offers a comparative view between our methodology and the other systems present in the competition. We must mention that the overall performance figures are measured using all the available MeSHes, not the pruned subset.

4 Result ranking

For this we take each lexical component of the key set of data extracted from the corpus and we try to find if the classified documents from the corpus approximate to possible synonyms of lexical component. For each lexical component of the key set of data extracted from the question, we calculated a list of lexical elements that can be considered similar in meaning using "cosine similarity" computed over distributed word representations (Mikolov et al., 2013). The vectors (100-dimensional) were computed using the word2vec[5] tool on a specific subset of Wikipedia combined with additional raw text resources provided as part of the BioASQ challenge. In order to compile the subset from Wikipedia we followed a simple boot-strapping procedure:

1. We downloaded the latest Wikipedia XML Dump at that date from the official web-site, on which we run a version of WikipediaExtractor[6], that was modified to preserve categories;

2. We seeded a list of categories, using the first level of categories on the Wikipedia site for the "Biomedical" main category;

3. We iterated 3 times through the entire corpus and we consolidated our category list, by adding categories that were associated with our initial category list, each time updating our seeded list;

4. We kept all documents that had at least one category from our final category list.

Given a "question" our IR process is: (a) we extract a list of keywords from the query, by removing function words from using a predefined dictionary; (b) we use the keywords to retrieve the top 1M documents from the initial corpus; (c) we re-rank our results and obtain a list with the top-10 most relevant documents. Document ranking

---

[5]https://github.com/dav/word2vec - accessed 2017-04-05
[6]https://github.com/bwbaugh/wikipedia-extractor - accessed 2017-01-28

Table 4: Test-set results

System Name	Mean precision	Recall	F-measure	Map	GMAP
Top 100 Baseline	0.2460	0.2845	0.1333	0.1606	0.0028
Top 50 Baseline	0.2470	0.2591	0.1920	0.1503	0.0024
fdu_5b	0.1865	0.2228	0.1791	0.1300	0.0084
Our System	**0.4000**	0.2222	**0.2857**	0.1238	**0.1238**
MCTeamMM	0.2266	0.1481	0.1249	0.0892	0.0005
MCTeamMM10	0.0326	0.1481	0.0436	0.0892	0.0005
Wishart-S1	0.0465	0.0484	0.0350	0.0237	0.0001

is performed using Equation 3, which is designed to take into account keyword synonymic coverage, but currently ignores synonymic frequencies in the text (in our empirical experiments we found that introducing this factor decrease the overall precision of the system - in our opinion, mainly because word-embeddings are prone to capturing contextual similarities, rather than actual synonymic behavior).

$$S_d = \frac{1}{k} \cdot \sum_{i=1}^{k} \max_{j=1}^{m}(cos(t_i, d_j)) \qquad (3)$$

where
S_d - is the relevance of document d
k - is the number of keywords in the query
m - is the number of words in the document
t_i - is the word embedding for term i in the query
d_j - is the word embedding for term j in the document

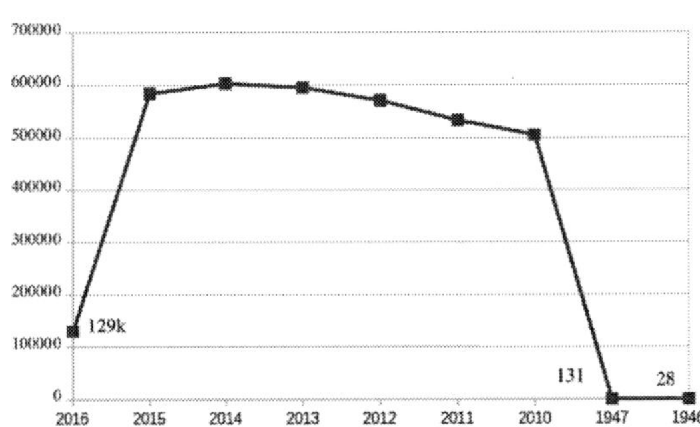

Figure 1: Distribution of publications each year

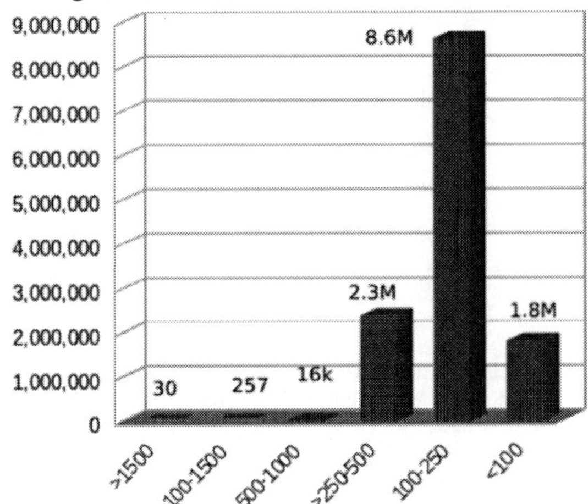

Figure 2: Distribution of words in articles

Table 4 shows the precision, recall and F-score of our system, measured on one of the datasets. It also offers a comparative view between our methodology and the other systems present in the competition. We must mention that the overall performance figures are measured using all the available MeSHes, not the pruned subset.

5 Snippets

Usually not all the text in the retrieved abstract is part of a good answer to a given question. So finding the most relevant, shortest part of the abstract was nest step.

To approximate the shortest span of text in each abstract of the documents, that represents the best response to the question, we selected a list of all the lexical tokens in the abstract text that correspond or might have generated the relevant label. At first glance, the snippet would be starting from the beginning of the first sentence that contains a

5

token from the list and finishing at the end of the last sentence that contains a token from the list.

Of course this list has a high probability of having duplicates. These duplicates have no value for detecting the shortest relevant text. So we calculate from the current abstract, the shortest span of text that still contains all of the lexical tokens but we ignore any duplicates in the list.

To help explain the previous statement we will use the following example:

```
"document": .... [token_1]....[token_1
    ]...[token_2].....[token_1]....[
    token_3].....[token_4].....[token_5
    ]....[token_1] ...
```

It can easily be seen in the example that the first iteration of "token_1" holds no value for the purpose of finding the shortest relevant span of text an neither does the second iteration even though it is position in closer proximity to another token from the list. The list is not in any way ordered so the placement of the second token: "token_2" in front of the first token "token_1" is irrelevant. The existence of a different token in front of the current token: "token_2" before "token_1" only means that this iteration of "token_1" is a viable candidate for the shortest relevant span of text. Finally the final iteration of "token_1" has no other tokens placed after it so we considered this iteration to hold less value for a snippet. No other token had a duplicate in this example so in this case the shortest most relevant span of text was:

```
"snippet": [token_2].....[token_1]....[
    token_3].....[token_4].....[token_5]
```

It is worth noting that there were of course cases when the system would present the snippet as being the same as the entirety of the abstract text.

6 Conclusions and future work

In this article we presented a "biomedical" oriented system that automatically assigns MeSH labels to documents in a large-scale corpus. Our apprach is based on a linear classifier, trained in a 1-vs-all style for each possible MESH.

The system then retrieves answers from said corpus for questions relevant to the medical field. Each question yields a number of "n" best ranked documents that relate to the question. We achieve this by first selecting the relevant lexical tokens from the questions. Then we use Word2Vec for 100 length vectors in order to calculate the cosine similarity to approximate "x" closest lexical concepts for each of the tokens from the question.

Our system also provides a corresponding list of "n" snippets from the best ranked documents, the shortest span of text which contain the information from the abstract most relevant for the current question. This is done by discarding any sentence from the abstract text that does not contain any token from a determined list or only contains low relevance duplicates of tokens from said list.

Currently we do not deal with determining and extracting lexical dependencies between words and we only focus on relevant-document retrieval. However, our future development plans include extending our system to be able to answer yes/no, factoid and item-list questions. Additionally we plan to include multilingual data from various sources and investigate cross-lingual techniques for document retrieval and machine translation for delivering the cross-lingual results in the user's native language.

References

L Douglas Baker and Andrew Kachites McCallum. 1998. Distributional clustering of words for text classification. In *Proceedings of the 21st annual international ACM SIGIR conference on Research and development in information retrieval*, pages 96–103. ACM.

Ronan Collobert and Jason Weston. 2008. A unified architecture for natural language processing: Deep neural networks with multitask learning. In *Proceedings of the 25th international conference on Machine learning*, pages 160–167. ACM.

Alaa M El-Halees. 2015. Arabic text classification using maximum entropy. *IUG Journal of Natural Studies*, 15(1).

Eui-Hong Sam Han, George Karypis, and Vipin Kumar. 2001. Text categorization using weight adjusted k-nearest neighbor classification. In *Pacific-asia conference on knowledge discovery and data mining*, pages 53–65. Springer.

Thorsten Joachims. 1998. Text categorization with support vector machines: Learning with many relevant features. *Machine learning: ECML-98*, pages 137–142.

Siwei Lai, Liheng Xu, Kang Liu, and Jun Zhao. 2015. Recurrent convolutional neural networks for text classification. In *AAAI*, volume 333, pages 2267–2273.

Bjornar Larsen and Chinatsu Aone. 1999. Fast and effective text mining using linear-time document clustering. In *Proceedings of the fifth ACM SIGKDD international conference on Knowledge discovery and data mining*, pages 16–22. ACM.

David D Lewis and Marc Ringuette. 1994. A comparison of two learning algorithms for text categorization. In *Third annual symposium on document analysis and information retrieval*, volume 33, pages 81–93.

David D Lewis, Robert E Schapire, James P Callan, and Ron Papka. 1996. Training algorithms for linear text classifiers. In *Proceedings of the 19th annual international ACM SIGIR conference on Research and development in information retrieval*, pages 298–306. ACM.

Larry M Manevitz and Malik Yousef. 2001. One-class svms for document classification. *Journal of Machine Learning Research*, 2(Dec):139–154.

A McCallum and K Nigam. 1998. A comparison of event models for naive bayes text classification; 1998. *Disponıvel em:¡ citeseer. nj. nec. com/mccallum98comparison. html.*

Tomas Mikolov, Ilya Sutskever, Kai Chen, Greg S Corrado, and Jeff Dean. 2013. Distributed representations of words and phrases and their compositionality. In *Advances in neural information processing systems*, pages 3111–3119.

Adwait Ratnaparkhi. 1998. *Maximum entropy models for natural language ambiguity resolution.* Ph.D. thesis, University of Pennsylvania.

FB Rogers. 1963. Medical subject headings. *Bulletin of the Medical Library Association*, 51:114–116.

Michael Steinbach, George Karypis, Vipin Kumar, et al. 2000. A comparison of document clustering techniques. In *KDD workshop on text mining*, volume 400, pages 525–526. Boston.

George Tsatsaronis, Michael Schroeder, Georgios Paliouras, Yannis Almirantis, Ion Androutsopoulos, Eric Gaussier, Patrick Gallinari, Thierry Artieres, Michael R Alvers, Matthias Zschunke, et al. 2012. Bioasq: A challenge on large-scale biomedical semantic indexing and question answering. In *AAAI fall symposium: Information retrieval and knowledge discovery in biomedical text.*

Min-Ling Zhang and Zhi-Hua Zhou. 2006. Multilabel neural networks with applications to functional genomics and text categorization. *IEEE transactions on Knowledge and Data Engineering*, 18(10):1338–1351.

Adapting the TTL Romanian POS Tagger to the Biomedical Domain

Maria Mitrofan
Research Institute for AI
"Mihai Drăgănescu"
Romanian Academy
13 "Calea 13 Septembrie",
Bucharest 050711, Romania
maria@racai.ro

Radu Ion
Research Institute for AI
"Mihai Drăgănescu"
Romanian Academy
13 "Calea 13 Septembrie",
Bucharest 050711, Romania
radu@racai.ro

Abstract

This paper presents the adaptation of the Hidden Markov Models-based TTL part-of-speech tagger to the biomedical domain. TTL is a text processing platform that performs sentence splitting, tokenization, POS tagging, chunking and Named Entity Recognition (NER) for a number of languages, including Romanian. The POS tagging accuracy obtained by the TTL POS tagger exceeds 97% when TTL's baseline model is updated with training information from a Romanian biomedical corpus. This corpus is developed in the context of the CoRoLa (a reference corpus for the contemporary Romanian language) project. Informative description and statistics of the Romanian biomedical corpus are also provided.

1 Introduction

Natural Language Processing (NLP) is one of the key technologies that can be employed to extract valuable information from unstructured text (e.g. discharge summaries, clinical notes, medical reference books, research papers, medical blog posts) and transform it into a desired form to support activities related to the healthcare domain.

NLP technologies have been adapted to the biomedical domain and applied on a vast amount of clinical data to enhance the research process and to extract relevant information from textual data. For example, clinical notes have been used for identifying cardiovascular risk factors (Abdulrahman and Meystre, 2015), electronic medical records have been used for detecting diabetes mellitus (Chung-Il et al., 2017). Jackson et al. (2017) applied NLP to extract symptoms of severe mental illness from clinical text. NLP tools have been proven to be an efficient way to enhance the identification on Alzheimer's disease (Shibata et al., 2016) and even the Human Genome project used NLP techniques in order to explore the relationships between biomedical literature and genes sequences (Yandell and W. H. Majoros, 2002).

A typical NLP pipeline consists in sentence delimitation, tokenization, part-of-speech (POS) tagging, lemmatization and parsing. More advanced NLP pipelines will perform NER and/or word sense disambiguation.

POS tagging (the process of labeling a token with a part of speech tag) is one of the initial pipelined components and it is an important step that performs morphosyntactic disambiguation. Therefore, the quality of the POS tagging is very important because cascading errors generated in POS tagging processes affect the overall performance of NLP pipelines. Consequently, it is very important that a POS tagger performs as optimally as possible.

The accuracy of a POS tagger is expected to be high (e.g. at least 97% for English) when the tagged text is similar, domain-wise, to the tagger's training data, but when the tagger is used on texts belonging to significantly different domains than the ones the tagger was trained on (e.g. train on newspaper articles and test on biomedical documents), its performance can degrade significantly. Ferraro et al. (2013) showed that the accuracy of state-of-the-art English POS taggers trained on news texts plummeted from 97% to 85% when POS tagging has been applied to clinical narratives, mainly because biomedical texts have different linguistic characteristics. Therefore the target domain adaptation of the POS tagger is needed.

Ferraro et al. (2013) note that there are multiple POS tagger domain adaptation techniques, out of which the simplest one is what they call "source-target labeled data aggregation" which refers to the

Proceedings of the Biomedical NLP Workshop associated with RANLP 2017, pages 8–14,
Varna, Bulgaria, 8 September 2017.

training of a POS tagging model based on a labeled corpus obtained from both the source and the target domains. A simplified version of this approach is the approach we follow here in order to adapt the baseline model of the trigram HMM-based TTL POS tagger (Ion, 2007) to Romanian biomedical POS tagging.

To obtain good POS tagging results with the source-target labeled data aggregation domain adaptation method, high-quality training data in both the source and the target domains is vital for a good performance of the POS tagger. Consequently most of the work concentrates on building high-quality training corpora, which are typically hand-made and slow to produce and which are, for this reason, very hard to find. In general, the lack of sufficient data in biomedical domain remains a barrier for biomedical NLP, especially for under-resources languages. Even though at the international level biomedical resources have been developed (e.g. HIMERA - a collection of historical medical documents manually annotated at semantic level with information relevant for public health, BMC - a corpus which contains full medical articles provided by BioMed Central, GENIA - a collection of 2000 biomedical abstracts annotated at syntactic and semantic level), at a national level (at least in Romania) it is very difficult to obtain texts for specialized corpora in the biomedical domain due to copyright laws and lack of biomedical literature published in Romanian language that is readily available in electronic format.

Efforts to improve the availability of Romainian biomedical training data for POS tagging are currently carried on. The most important is the Co-RoLa project which was started in 2012 by the Romanian Academy Research Institute for Artificial Intelligence "Mihai Drăgănescu" (RACAI) and the Institute for Computer Science in Iași. It aims to create a reference corpus of the contemporary Romanian language (CoRoLa) (Mititelu et al., 2014), which will be useful for different types of NLP tasks, including POS tagging.

In what follows, we will briefly review related word in POS tagging domain adaptation for the biomedical domain (Section 2), we will introduce the Romanian biomedical corpus that we used to adapt TTL to the biomedical domain (Section 3), we will briefly describe TTL (Section 4) and we will present our initial experiment in Romanian biomedical POS tagging (Section 5). The paper ends with our concluding remarks (Section 7).

2 Related Work

Domain adaptation received significant attention from the NLP research community and multiple approaches have been developed to improve the tagging accuracy and to reduce the errors caused by out-of-vocabulary words. A very common approach used for domain adaptation is to combine both the source and the target training data to train a new model. This method was used by Coden et al. (2005) when an HMM POS tagger was trained on both news and a medical corpus of clinical notes. After this experiment they reported an accuracy of almost 93% when the tagger was tested on the medical test set, compared to a little over 87% when the tagger was trained on the news corpus and tested on the medical test set.

For the GENIA POS tagger, Tsuruoka et al. (2005) presented several experimental results for domain adaptation on GENIA, PennBioIE and Wall Street Journal (WSJ) corpora. POS tagging performances has been evaluated for seven different combinations of the corpora as the training data. When the tagger was trained on WSJ corpus (without the distinction between nouns and proper nouns) and tested also on a test set extracted from WSJ corpus (in-domain testing), the accuracy was 97.20%, but when the tagger was applied on test sets extracted from biomedical corpora (out-of-domain testing), the accuracy dropped significantly: 91.55% on GENIA and 90.51% on PennBioIE. On the other hand, when the GENIA tagger was trained both on WSJ and GENIA corpora, it achieved an accuracy of 98.32% on the GENIA test set and an accuracy of 96.96% on the WSJ test set (and a lower accuracy on PennBioIE test set, 91.98%). This shows that domain adaptation is worth doing even though in-domain accuracy may drop a little.

cTAKES tagger is an example of a biomedical tagger that demonstrates the variability of the biomedical domain. This tagger was trained with Mayo Clinic's notes and tested on a set of clinical notes from Kaiser Permanente Southern California (KPSC) on which it obtained an accuracy of 88.1%. Moreover, the cTAKES tagger tested on set of clinical notes from the University of Pittsburg Medical Center (UPMC) achieved an accuracy of 88.3%. This is to show that POS tagging in the biomedical domain is more difficult than, e.g.

news POS tagging, mainly because of the extensive lexicon of the domain.

Finally, we present two experiments demonstrating that good accuracy can be obtained with in-domain biomedical data even with small training sets. Smith et al. (2004) trained the MedPost tagger on 5,716 manually tagged sentences taken from Medline abstracts within the Genomics domain and achieved an accuracy of 97.43% on 1000 sentence test set extracted also from MEDLINE abstracts. In order to train the Brill tagger on biomedical domain, Campbell and Johnson (2001) tagged by hand 100,000 words from a corpus of discharge summaries, 90% of the hand tagged corpus was used to train the tagger an the remaining 10% was used to test the tagger. This process was repeated ten times and achieved an accuracy of 96.9%, each time using a different 10% as the test set.

3 Corpus Structure

In order to perform domain adaptation we have developed a domain-specific training corpus, because sub-domain languages present distinct linguistic features, usually not found in general language, in this case Romanian language.

The process of collecting the texts was not an easy task, firstly because of the intellectual property restrictions and secondly because in general, biomedical literature is published in English and not in the Romanian language. At the end of this process the Romanian medical corpus contained texts from different sources such as medical books published at the Romanian Academy Publishing House and Polirom publishing house, free medical online resources, medical blogs, online courses made for medical students.

The biomedical corpus has evolved from a collection of texts extracted from different biomedical sub-domains such as: cardiology, endocrinology, diabetes, oncology, surgery, genetics, nephrology, neurology, psychiatry etc. The textual resources available in the corpus were initially available in different formats such as .doc and unprotected .pdf and they had to be converted into a raw text format in order to be annotated by our processing tools (Tufiş et al., 2008). The conversion of the files involved a boilerplate removal step in which footers, headers, page numbers, figures, tables, footnotes, etc. have been removed. For this step we used the tool designed by (Moruz and Scutelnicu, 2014). In order to improve the linguistic annotation we considered only texts with correct diacritical characters, encoded in UTF-8.

The Romanian biomedical corpus used for domain adaptation of the TTL POS tagger contains about 206,020 sentences and 4,390,707 million tokens (words and punctuation) distributed in more than nine medical sub-domains (see above) extracted from academic books and journals and one which contains information from different free medical online resources such as medical blogs and Romanian medical publications.

The resources extracted from online sources have not been grouped into medical categories because most of them belong to more than one medical category and medical expertise was needed in order to fulfill this task. Furthermore the POS tagging step is not affected by this lack of classification. All the texts were split into tokens, POS tagged and lemmatized with the baseline model of TTL (see Section 5).

Table 1 shows some statistics of the automatically POS tagged biomedical corpus: we counted all tokens (words plus punctuation), words (functional words and content words), unique lemmas and sentences. Content words also included abbreviations because these represent an important feature of the biomedical texts. The punctuation count is obtained by subtracting the words count from the tokens count (Table 2). From a statistical point of view, the corpus is balanced in terms of tokens per sentence, content words per sentence and punctuation per sentence (Table 2 and Table 3) when comparing sub-domains.

Table 3 shows that the texts obtained from online resources contain the highest use of content words per sentence; at the other end the texts from endocrinology domain use the lowest number of content words. An interesting fact is that the average number of punctuation per sentence contained in the texts extracted from online sources remains in compliance with the average number of punctuation used in academic medical literature.

In Table 4 the distribution of content words is presented among the POS tags types. While online resources texts make use of more nouns and less adjectives, the other medical sub-domains use less nouns and more adjectives. A characteristic specific to the biomedical domain, which it is also shown in table 4 is represented by the high use of the total nouns and adjectives.

# tokens, punctuation included	4,390,707
# words	3,750,242
# unique lemmas	101,348
# sentences	206,020
average tokens per sentence	21.31
average words per sentence	18.20
average punctuation per sentence	3.10

Table 1: Statistics over the Romanian medical corpus.

4 Tokenizing, Tagging and Lemmatizing (TTL) Platform

TTL is a Perl module supporting Romanian, English, French and Bulgarian, with the following functionalities: sentence splitting, tokenization, POS tagging, lemmatization, chunking and Named Entity Recognition (NER).

TTL's tokenizer takes two input parameters (the code of the language and the sentence) and returns a list of tokens. Moreover the tokenization procedure is language independent and identifies clitics, contractions and multiword expressions (MWEs), provided that language-dependent resources exist (i.e. list of MWEs and affix words that should be split).

The POS tagger is a heavily-improved reimplementation of the Hidden Markov Models (HMM) tagger presented in Brants (2000). It uses the tiered tagging technology (Tufiș, 1999; Ceaușu, 2006) for a more accurate POS labeling with a large tagset: the MSD tagset [1]. The Romanian MSD tagset has 736 labels and the general purpose Romanian language POS tagging accuracy is over 98% with this tagset (Tufiș, 1999).

Lemmatization is achieved after the POS tagging process is complete. TTL lemmatizer uses a large human-validated Romanian inflected lexicon, currently holding 1,152,506 entries. For the out-of-dictionary words, the TTL lemmatizer selects the most probable lemma provided by a five-gram letter Markov Model-based guesser (see Ion (2007) for details).

Chunking is another functionality of the TTL platform and it is based on a set of regular expressions applied on sequences of POS tags. The TTL chunker recognizes nominal, verbal, adjectival, adverbial and prepositional phrases.

5 Adapting TTL to the Biomedical Domain

As already stated in the Introduction, we attempted to adapt the baseline model of the Romanian TTL POS tagger to the biomedical domain by following the "source-target labeled data aggregation" paradigm. In our case, we have updated the baseline model's parameters by training on a sample of the Romanian biomedical corpus, for reasons to be explained below.

It is a well-known fact that the performance of a POS tagger depends crucially on the quality of the labeled corpus on which it trains. Thus, the baseline model for Romanian POS tagging that TTL uses is based on training on news (some "Adevărul" and "România Liberă" issues, 98,194 tokens) and fiction (Orwell's "1984", 118,357 tokens) corpora whose POS labeling was carefully checked by trained linguists, word by word (Tufiș, 2000).

Our initial experiment in biomedical POS tagging domain adaptation focused on experimentally verifying the assumption that we can get good results with an in-domain corpus whose POS labeling *is semi-automatically corrected*. That is, what results do we get if the biomedical corpus that is used to adapt TTL to the domain is not checked word for word but is corrected using some semi-automatic procedures (to be described below) whose output is checked by the trained linguist.

Since we could not hope to manually check 4.4M tokens as our Romanian biomedical corpus has (nor did we want to commit to such a task), we performed a random sampling of that corpus in order to obtain reasonable-sized train and test corpora. We concluded that, with our resources, we could check around 600K tokens, which, according to the English domain adaptation literature cited above, is a reasonable size. Thus, after splitting our sample into train and test sets, the train set contained 545,977 tokens (words and punctuation) and the test set contained 60,520 tokens, which is about 10% of the part we selected. The selection was done randomly, but enforcing the following conditions:

- We have sentences of all lengths from the Romanian biomedical corpus (short, average and long);

- All sentences have Romanian diacritics in

	Sentences	Tokens	Content words	Punctuation
Online resources	52,708	1,146,052	772,564	151,189
Cardiology	35,505	754,394	418,619	110,850
Surgery	51,367	989,335	550,037	156,140
Diabetes	33,538	775,017	411,393	114,123
Oncology	22,746	523,568	281,331	78,693
Endocrinology	10,156	202,341	112,826	29,470
Total	206,020	4,390,707	2,546,770	640,465

Table 2: Statistics on medical domains

	Sentences	Tokens	Content	Punctuation
Online resources	52,708	21.74	14.65	2.86
Cardiology	35,505	21.24	11.79	3.21
Surgery	51,367	19.26	10.65	3.03
Diabetes	33,538	23.10	12.26	4.44
Oncology	22,746	23.01	12.36	3.45
Endocrinology	10,156	19.92	11.10	2.90
Total	206,020	21.31	12.34	3.10

Table 3:   The average number of tokens, content words and punctuation per sentence by biomedical subdomain

place and are written using the Romanian Academy Romanian writing reform (i.e. using 'â' instead of 'î' inside words);

- There are no duplicate sentences.

Both the train and the test sets were automatically POS tagged with the TTL's baseline model. The test set was manually checked, word by word, by a trained linguist. The manual correction procedure involved reading each sentence from the test set, word by word, and making sure that the POS labellings are correct (the test set had to be thoroughly checked because the POS tagger performance was going to be measured against it).

For the train set, to speed up the correction process, we adopted the following semi-automatic approach:

- We extracted the list of unknown words with all their inflected forms (7,816 unique word forms) and checked their POS labellings, adding alternate analyses where it was necessary (e.g. adding a noun analysis for an existing adjective analysis);

- Noticing that TTL does not (usually) assign the wrong POS to a word (e.g. if a word is a noun, TTL will recognize it as such but, for unknown words, it may give the wrong

gender or case), we automatically replaced the POS labels of all unknown words in the train set with the corresponding POS labels from the curated unknown list. We were thus able to automatically fix 26,184 occurrences of unknown words in the train set;

- We built a TTL POS tagging model only from the train set and re-tagged the train set with it (we call this a 'biased evaluation'). We then inspected manually all the differences in POS labeling between the original tagging and the biased tagging. Some more (about 2% of the train set) inconsistencies were fixed this way;

- We also corrected every error that we saw in the train set, *but without going through it, word by word*.

Tables 5 and 6 present the TTL POS tagger accuracy on the biomedical test set. From Table 5 we see that general POS tagging accuracy degrades a little and this can be explained by the fact that the biomedical train set is not yet fully correct when it comes to POS labeling.

The baseline TTL model is trained over texts that were corrected *at word-level* by trained linguists while our biomedical train set was mostly automatically corrected with only a small part being manually validated. That the biomedical train

	Nouns	Verbs	Adjectives	Adverbs	Abbreviations
Online resources	477,208	137,729	120,382	24,717	12,528
Cardiology	224,758	70,684	102,039	14,717	6,421
Surgery	284,549	100,567	138,777	19,695	6,449
Diabetes	222,905	82,638	81,327	17,559	6,964
Oncology	153,955	53,357	58,350	9,528	6,141
Endocrinology	59,776	21,074	25,262	4,601	2,113
Total	1,423,151	466,049	526,137	90,817	40,616

Table 4: POS statistics for content words in each biomedical sub-domain.

	Errors	Accuracy
Baseline model	1,068	98.23%
Biomedical model	1,310	97.83%

Table 5: Overall TTL accuracy on the test set

	Errors	Percent
Baseline model	486	45.50%
Biomedical model	448	34.19%

Table 6: Errors on biomedical terminology

set still contains general language POS annotation errors becomes evident when the most frequent errors (on the test set) are identified (which *are not produced* by the baseline model):

- Verb 'a fi' (English 'to be') can occur as an auxiliary ('a fi' plus past participle) or main (61 errors);

- Verb 'a avea' (English 'to have') can also occur as an auxiliary (when forming the present perfect tense) or main (15 errors).

Table 6 shows the benefit of doing domain adaptation, *even with a minimally corrected in-domain corpus*: the percentage of errors relating to biomedical terminology (i.e. nouns, main verbs, adjectives and adverbs that are specific to the domain) is smaller when we use the adapted POS tagging model. At this point, if the degradation in general-purpose POS tagging is acceptable (0.4% in our case) the much lower error rate (11.31% in our case) in biomedical terminology POS tagging could be of help in applications such as biomedical NER.

6 The Availability of the Data

After the train set and the test set will be checked in detail ("word by word") both of them will be freely available for download[2] and non-commercial use. Special use-cases require license permissions from the author.

The biomedical corpus will be available in the context of the CoRoLa project copyright agreement signed with the publishing houses and with the editorial offices representatives. The whole corpus will be available to the public through KorAP platform (Banskiand et al., 2013), but will not be downloadable. The KorAP platform allows multiple linguistic types of searches in the corpus. However, all the results of the interrogation of the corpus outside the scope of the copyright restrictions will be downloadable.

7 Conclusions and Future Work

This paper presents a newly created text corpus aimed at providing support for NLP on biomedical text and an initial experiment about the adaptation of the TTL POS tagger to the biomedical domain. Currently our text corpus is still under development, but the available data and the biomedical TTL POS tagger can already be considered important resources in order to perform more advanced NLP tasks in the Romanian biomedical domain. To the best of our knowledge, the Romanian biomedical corpus is the first of its kind.

Our initial experiment was promising in the sense that, with minimal POS labeling correction efforts, we were able to improve the accuracy of the tagger where it matters most for other biomedical applications using POS tagging: the biomedical terminology. Thus, the error rate of biomedical terminology was reduced by 11.31%. We plan to fully validate the biomedical train set, with the help of trained linguists, and repeat the experiments to ensure that we obtain comparable (with the baseline) general language POS tagging accuracy (over 98% accuracy) while lowering even

---

[2]http://slp.racai.ro/index.php/resources/

more the error rate on biomedical terminology.

Compared to other corpora used for domain adaptation, our biomedical train set is larger (545,977 tokens) than most of the POS train sets. Another important characteristic of the biomedical corpus used for the adaptation of the TTL POS tagger is the variability of its lexicon: it contains words from five major biomedical sub-domains and a collection of texts extracted from online sources. Thus, we think that any POS tagger trained on it will perform better on a wider range of Romanian biomedical texts.

The train and test sets will also be annotated with biomedical named entities and parsed with our Romanian Universal Dependencies parser developed in the SSPR project (Mititelu et al., 2016). Thus, we will have a Romanian biomedical corpus that can be used as training data for other useful NLP tasks such as biomedical terminology identification, biomedical NER, biomedical text mining, etc.

References

Khalifa Abdulrahman and Stéphane Meystre. 2015. Adapting existing natural language processing resources for cardiovascular risk factors identification in clinical notes.

P. Banskiand, J. Bingel, N.Diewald, E. Frick, M. Hanl, M. Kupietz, P. Pezik, C. Schnober, and A. Witt. 2013. The new corpus analysis platform at ids mannheim.

T. Brants. 2000. Tnt: a statistical part-of-speech tagger. In *Proceedings of the sixth conference on Applied natural language processing*.

D. Campbell and S. Johnson. 2001. Comparing syntactic complexity in medical and non-medical corpora.

Alexandru Ceaușu. 2006. Maximum entropy tiered tagging. In *Proceedings of the 11th ESSLLI student session.*.

Wi Chung-Il, E.and Voge G. Sohn S.and Rolfes M. C.and Seabright A.and Ryu, and H Liu. 2017. Application of a natural language processing algorithm to asthma ascertainment: An automated chart review.

A. R. Coden, S. V. Pakhomov, R. K. Ando, P. H. Duffy, and C. G. Chute. 2005. Domain-specific language models and lexicons for tagging.

J.P. Ferraro, H. Daumé III, S. L. DuVall, W. W. Chapman, H. Harkema, and P. J. Haug. 2013. Improving performance of natural language processing part-of-speech tagging on clinical narratives through domain adaptation.

Radu Ion. 2007. *Word Sense Disambiguation Methods Applied to English and Romanian (in Romanian)*. Ph.D. thesis, Romanian Academy.

R. G. Jackson, R. Patel, N. Jayatilleke, A. Kolliakou, M. Ball, G. Gorrell, and R. Stewart. 2017. Natural language processing to extract symptoms of severe mental illness from clinical text: the clinical record interactive search comprehensive data extraction (cris-code) project.

V. Barbu Mititelu, E. Irimia, and D. Tufiș. 2014. Corola – the reference corpus of contemporary romanian language. In *Proceedings of the Ninth International Conference on Language Resources and Evaluation - LREC*. pages 1235–1239.

Verginica Barbu Mititelu, Radu Ion, Radu Simionescu, Andrei Scutelnicu, and Elena Irimia. 2016. Improving parsing using morpho-syntactic and semantic information, in revista romana de interactiune omcalculator.

Alex Moruz and Andrei Scutelnicu. 2014. An automatic system for improving boilerplate removal for romanian texts. In *Proceedings of the 10th International Conference "Linguistic resources and Tools for Processing the Romanian Language*.

D. Shibata, S. Wakamiya, E. Aramaki, and A. Kinoshita. 2016. Detecting japanese patients with alzheimer's disease based on word category frequencies.

L. Smith, T. Rindflesch, and W. J. Wilbur. 2004. Medpost: a part of speech tagger for biomedical text.

Y. Tsuruoka, Y.Tateishi, J. D. Kim, T. Ohta, J. McNaught, S. Ananiadou, and J. I. Tsujii. 2005. *Developing a robust part-of-speech tagger for biomedical text.*.

D. Tufiș, R. Ion, A. Ceaușu, and D. Ștefănescu. 2008. In proceedings of the 6th language resources and evaluation conference-lrec.

Dan Tufiș. 2000. Using a large set of eagles-compliant morpho-syntactic descriptors as a tagset for probabilistic tagging. In *Proceedings of LREC*.

Dan Tufiș. 1999. *Tiered tagging and combined language models classifiers.*. Springer.

M. D. Yandell and W. H W. H. Majoros. 2002. Genomics and natural language processing.

Discourse-Wide Extraction of Assay Frames from the Biological Literature

Dayne Freitag
SRI International
9988 Hibert Street, Suite 203
San Diego, CA 92131, USA
freitag@ai.sri.com

Paul Kalmar
SRI International
9988 Hibert Street, Suite 203
San Diego, CA 92131, USA
paul.kalmar@sri.com

Eric Yeh
SRI International
333 Ravenswood Avenue
Menlo Park, CA 94025, USA
yeh@ai.sri.com

Abstract

We consider the problem of populating multi-part knowledge frames from textual information distributed over multiple sentences in a document. We present a corpus constructed by aligning papers from the cellular signaling literature to a collection of approximately 50,000 reference frames curated by hand as part of a decade-long project. We present and evaluate two approaches to the challenging problem of reconstructing these frames, which formalize biological assays described in the literature. One approach is based on classifying candidate records nominated by sentence-local entity co-occurrence. In the second approach, we introduce a novel virtual register machine that traverses an article and generates frames, trained on our reference data. Our evaluations provide evidence that best performance in the task ultimately hinges on an integration of information distributed over multiple sentences.

1 Introduction

Biological event and relation extraction have been the focus of considerable study in recent years, resulting in the availability of annotated corpora (Kim et al., 2003; Pyysalo et al., 2007; Kim et al., 2008; Thompson et al., 2009). In the interest of replicability and progress on critical challenges, such resources typically decompose the hard problem of factual understanding into several simpler problems, such as entity recognition, binary relation detection, and co-reference resolution.

This methodology is subject to several criticisms. The reliance on thorough annotation imposes overheads that prevent rapid progress. The targeting of a fixed set of simplified, typically binary relations does justice neither to the complexity of information expressed in a typical sentence, nor to the biological processes under discussion. And the methodology places an emphasis on pieces of information amenable to expression in individual sentences, leaving untouched information that can be assembled only through traversal of multiple sentences.

In this paper we address the problem of constructing multi-slot knowledge frames from the technical literature on cellular signaling networks. The frames in our study are a faithful representation of assays reported in this literature, called *datums*, with only approximate localization to specific textual regions. We have no one-to-one mapping between frames and sentences, no guarantee that the slots of a frame co-occur in a single sentence, and no universal presentational convention governing the sequence of slot-relevant expressions. Nevertheless, we seek to learn procedures for populating frames in new documents.

Success in this endeavor would have significant practical impact. If we can automate the separation of experimental evidence from common knowledge and speculation, we have the means to construct a high-quality biomedical resource of use to both experimental and computational biologists. Our efforts, for example, ultimately seek to automate the maintenance and extension of high-fidelity machine models of signaling pathways associated with Ras-driven human cancer.

We offer three contributions. First, we describe a problem of clear biomedical significance that involves synthesis of information distributed across a document, one that poses pertinent challenges to the current practice of machine reading. Second, we describe and evaluate an approach (the *frame classification* approach) that formalizes this problem as a binary classification of frames nominated

Proceedings of the Biomedical NLP Workshop associated with RANLP 2017, pages 15–23,
Varna, Bulgaria, 8 September 2017.

by protein pairs co-occurring in sentences. We provide evidence that good performance on this problem requires attention to how entities are referenced across a document, even in multiple documents, not just in the nominating sentence. Finally, we describe and evaluate an approach (the *register machine* approach) that attempts to correct deficiencies of the frame classification approach, specifically its limiting reliance on sentence-local juxtaposition of frame slot elements. This approach formalizes the frame extraction problem as learning the best sequence of instructions for frame generation through document traversal.

2 Related Work

Progress in biomedical information extraction (BioIE) is measured against shared annotated corpora that decompose the problem into entity extraction and sentence-level relation and event detection (Kim et al., 2003; Pyysalo et al., 2007; Kim et al., 2008; Thompson et al., 2009). The structure of these tasks has remained remarkably stable over the years, differing in some important ways from the task addressed in this paper. Most notably, the canonical BioIE task is highly localized and mostly agnostic to discourse context. The objective is to determine whether a single sentence expresses some event of interest–gene expression, phosphorylation, regulation, etc.–and, if so, what roles the entities appearing in the sentence play in the putative event. The "events" detected in this fashion are divorced from their discourse context (modulo coreference resolution), although some attention has been paid to epistemic qualifications, such as negations and speculations (Kim et al., 2011). We have posed ourselves a more focused task—extract the experiments described in a paper—and are forced to do without reliable sentence-level annotation.

There is no doubt that our system must respond to some of the same expressions that are addressed in some of these shared tasks. In particular, the Genia Event Extraction Task (Kim et al., 2011) targets phosphorylation and regulation events involving phosphorylation, among other things. Many of these event mentions are encountered in sections detailing experiments. Thus, our task can be addressed in part through disambiguation and assimilation of these events—which were actually observed in experiments? In this paper, we describe an approach to datum extraction

that elaborates this idea.

Our focus on multi-slot, multi-sentence factual frames is reminiscent of early formulations of the information extraction problem used in the Message Understanding Conference (MUC) (Grishman and Sundheim, 1996; Chinchor et al., 1993). Over successive iterations of MUC, target frames became quite elaborate, similar in complexity to the datums we ultimately seek to populate. Many of the tasks that the information extraction community views as canonical, including named entity recognition, co-reference resolution, word sense disambiguation, and relation and event extraction, were introduced as simplifications of the core frame-filling task in MUC6. The field has since largely neglected the discourse-wide frame-filling challenge.

Of course, to take it up again and address it with the latest machine learning techniques, we require heuristics to align slot-level information found in reference frames to expressions in training sentences. Using such heuristics, in combination with structured ground-truth data, such as our collection of datums, is commonly referred to as *distant supervision*, an approach pioneered on a biomedical extraction problem (Craven and Kumlien, 1999). Relatively little work has applied distant supervision to discourse-level extraction problems. A counter-example is Reschke et al. (2014), which addresses event extraction at the document level, attempting to populate event-related Wikipedia "info boxes" with article source text. The Reschke et al. approach employs SEARN (Daume III et al., 2009), a technique that reduces complex structured classification problems into simpler sequence learning problems, finding that it yields performance superior to several strong baselines.

Recently, the problem of understanding accounts of experiments in the biological literature has been the focus of a small amount of study (Dasigi et al., 2017; Burns et al., 2016). This work, which springs from the same motivation and shares some of the same data as our own, is largely concerned with modeling the discourse structure of experimental narratives. It is therefore largely complementary to our work, which targets factual experimental details. Success in discourse modeling promises to solve key problems that we face, such as the segmentation of the text into distinct experiments.

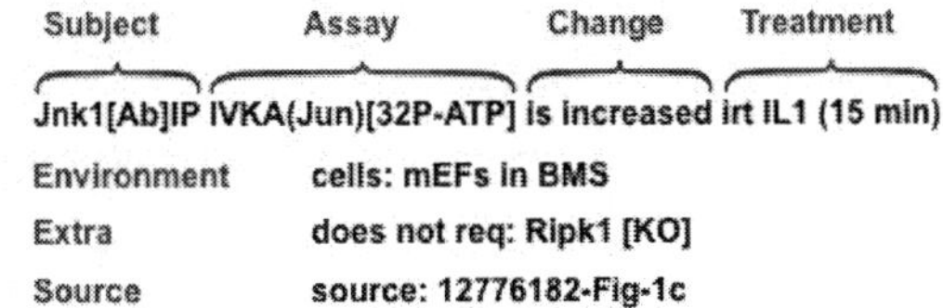

Figure 1: An annotated Pathway Logic datum.

3 Problem

The Pathway Logic (PL) project pursues high-fidelity signaling pathway models centering on the Ras family of proteins (Eker et al., 2004). Part of the effort involves a manual curation of experimental results, which has resulted in approximately 50K records, each containing a detailed formal representation of a reported experiment and its outcomes. Such records, called *datums*, retain pointers to the papers and figures from which they were derived.

Figure 1 displays a typical datum in its compact formal syntax, highlighting the four key components: the *assay*, encoding the type of assay conducted (here, an in-vitro kinase activity assay); the *subject*, the entity whose response was measured ("Jnk1"); the *treatment*, the substance applied to the cellular environment (here, some member of the IL-1 family, either IL-1 alpha or IL-1 beta); and the *change* or experimental outcome. It should be apparent from the figure that the typical datum records many additional experimental details. We refer to the combination of these four fields, stripped of such qualifiers, as a *simplified datum*, and seek to reconstruct these 4-tuples in our experiments.

Notable among the fields in Figure 1, is an encoding of the source of the datum, most frequently as a PubMed ID and figure reference. The datum curator, not a computational linguist, found it most natural to localize datums to the figures displaying assay outcomes. As a consequence, we do not have access to a simple procedure for identifying specific textual expressions for the various datum elements. In fact, the data comes with no guarantee that such expressions are present at all.

However, after a manual review of a large number of datums, we know that while some datums are not adequately described in the text of annotated articles, most are. Furthermore, the alignment of datums to figures enables weak localization of datum elements to individual sentences, be-

cause figure captions and body sentences containing figure references are on average relatively rich in information needed to populate the simplified datums attached to corresponding figures.

> 1. *Phosphorylation and activation of JAK1 and Stat6 are essential for induction of Stat6 DNA binding activity.* 2. *To ascertain whether the decrease in Stat6 DNA binding activity in the SOCS-1 stable transfectants was due to inhibition of JAK1 kinase activity, we immunoprecipitated lysates from cells untreated or treated with IL-4 with Abs to JAK1 or Stat6 and probed with Ab to phospho-tyrosine.* 3. *Induction of JAK1 and Stat6 phosphorylation in the SOCS-1 stable clones was reduced when compared with control (Fig. 3A), while induction in the SOCS-2 stable clones (Fig. 3B) and in the SOCS-3 stable clones (Fig. 3C) was similar to that of controls.* 4. *To further confirm that SOCS-1 suppresses JAK1 activation, we measured the IL-4-induced kinase activity of JAK1 in the SOCS-1 stable clones by in vitro kinase assay.*

Table 1: Example sentences potentially expressing the key elements of a "phos" datum.

Table 1, which excerpts four contiguous sentences (we have numbered them for convenience) from a relevant article, renders this concrete, but also illustrates some of the subtleties involved. Our data notes two distinct phosphorylation assays in this passage, both linked to Sentence 3 (the only sentence with figure references), corresponding to the subjects JAK1 and Stat6, respectively, each of which are phosphorylated (i.e., the change of the "phos" assay is "increased") in response to IL-4 (the treatment).

This passage is abundant in evidence about the relevant experiments, but the information is distributed. Sentences 1 and 3 both contain inflections of "phosphorylate," providing evidence that a "phos" assay was conducted, but both lack the treatment IL-4, which is referenced in Sentences 2 and 4. Note that the "phosphorylate" sentences are rich in entities, potentially posing a combinatoric discrimination problem. Ultimately, if we wish to extract the two target datums at Sentence 3, information about the experimental treatment must be pulled in from one of the adjacent sentences, and we must determine that exactly two datums are warranted.[1]

---

[1] Actually, a number of experimental variants are under discussion in this passage. These are captured the database in supplementary records called "extras." Extras are not the

Whatever the textual evidence for datums in a paper, our problem is essentially extraction at the level of documents. Formally, we are given a set of examples $\{\langle d_i, y_i \rangle\}$, in which d_i is a document and y_i is a set of tuples $\{\langle s_{ij}, t_{ij}, a_{ij}, c_{ij} \rangle\}$, the elements of each tuple representing subjects, treatments, assay types, and observed changes, respectively. Assay and change values are drawn from closed classes, assays from the set of types represented in the Pathway Logic knowledge base, and changes from the set {*increased*, *decreased*, *unchanged*}. Subjects and treatments are drawn from the effectively open class of chemicals used for experiments in the literature. In practice, they are usually proteins, and in our experiments these two slots take Uniprot IDs. Our extraction task involves inferring the correct set y_i, given some d_i.

4   Approaches

We investigate two distinct approaches to this problem, the *frame classification* approach and the *register machine* approach. The first is applicable only to subject-treatment pairs that co-occur in individual sentences, while the second approach can in principle associate subjects and treatments found in different sentences.

4.1   Frame Classification

We observe that datum subjects and treatments tend to be mentioned together in individual sentences. This motivates a simple framing of the datum extraction problem as binary classification. Specifically, if we fix the assay type (e.g., "phos") and change (e.g., "increased"), we can view each document as a set of co-mentioned proteins—all pairs of proteins mentioned together in some sentence—and attempt to distinguish pairs in the subject-treatment relation from other pairs. Of course, we must perform this procedure for all assay-change pairs of interest.

We follow an approach to featurization proposed in Xu et al (2016). Consider the set of sentences containing protein entities P_1 and P_2. Given a target assay-change configuration we train *two* binary classification models, one to distinguish cases in which P_1 and P_2 are subject and treatment, respectively, and one for the opposite assignment. Our feature vectors have four parts, each part containing features that require the frequency of lexical unigrams and bigrams found in

various sentence contexts. Thus, the word "protein" corresponds to three distinct features: one feature recording its frequency of occurrence before P_1 in the set of sentences, between P_1 and P_2 (encountered in that order), between P_2 and P_1 (encountered in that order), and after P_2, respectively.

We also included and recorded a small performance benefit from two non-lexical features. First, observing that datum protein pairs tend to be more frequent than others, we defined a feature that reflects the number of sentences in which a pair co-occurs. Second, we defined indicator features that reflect whether specific proteins fill a subject or treatment role anywhere in the training data.

Admittedly, this approach suffers from certain limitations, most obviously limited recall, as it can only distinguish datums whose subject and treatment co-occur in a sentence—e.g., discarding some 40% of phosphorylation datums. And as noted, because the classification problem is conditioned on assay and change, we must learn a separate classifier for each observed assay-change combination. This is tractable in practice, because the number of frequently observed assay-change combinations occurring is manageable.

4.2   Register Machine

To accommodate the distribution of relevant information across the sentences in a discourse, we imagine a model capable of traversing sentences, accumulating information, and synthesizing datums. We suppose that datums are produced by a virtual machine with four registers (one for each of the slots in a simplified datum) and two cursors (to traverse the sentences in a caption and article body, respectively). At each time step, the machine can execute an instruction to advance either cursor, populate or delete the contents of registers, or produce one or more datums. Specifically, we define the following instructions:

- **advance*Section*Cursor**, where *Section* can be either *Caption* or *Body*. One of the cursors is advanced to the next entity within the current sentence, if present, or to the beginning of the next sentence in body or among figure captions.

- **set*ClosedValue***, where *Closed* can be either *Assay* or *Change*, and *Value* is one of the legal values for the indicated closed-class register. The register becomes populated with

---

focus of this paper's work, but are ultimately important.

the specified value, replacing any previous contents.

- **set*Open*from*Section***, where *Open* is either *Subject* or *Treatment*. The indicated register is populated with the entity under the cursor for *Section*. This instruction is illegal if there is no such entity.

- **add*Open*from*Section***. This instruction is like the previous one, except the entity is accumulated into the indicated register. As this implies, open-class registers can hold multiple entities.

- **delete*Register*** empties the indicated register.

- **deleteAll** empties all registers.

- **produceDatums** causes datums to be generated from register contents. A different datum is generated for each distinct combination of entities in the subject and treatment registers.

Let us suppose we are given a sequence of instructions $I = i_1 \cdots i_m$ applying the machine to some example $\langle d_i, y_i \rangle$. It is easy to see than any such I yields a set of datums y_i^*, which we can formalize as some function, $F(d, I) = y$.[2] Further, we can speak of a policy $\pi(d) = I$ that nominates instructions sequences, given a document. Ultimately, our objective is to find the best policy:

$$\operatorname*{argmin}_{\pi} \sum_i L(F(d_i, \pi(d_i)), y_i) \qquad (1)$$

Here, $L(y^*, y)$ is the loss experienced by some machine-generated set of datums y^* with respect to the ground-truth y. In practice, we seek to optimize the F1 of extracted datums versus ground truth under a strict equality standard, i.e., only those datums that agree in all slots with some ground-truth datum are counted as successes.

Of course, Equation 1 is difficult to satisfy directly. Instead, we seek to learn a local ranking model for individual instructions. Let $S(d, I_{1,k})$ represent the state of the machine after executing k instructions $I_{1,k} = i_1 \cdots i_k$ against document d, including the positions of the cursors, the state of the registers, and any generated datums. We seek to learn a local policy $\hat{\pi}(S(d, I_{1,k})) = i_{k+1}$ that chooses the best next instruction.

---

[2]We posit that illegal instructions (e.g., advancing a cursor at the end of the document) have no effect.

Learning $\hat{\pi}$ is essentially a ranking problem: given all legal instructions in the current state, which is best to execute? We therefore adopt a learning-to-rank approach, training an empirical model to map machine states to real values, such that the highest-scoring instruction is the best to execute in the current state. To this end, we implemented an oracular policy (henceforth the "oracle") that nominates instructions based on full knowledge of ground truth. Given our uncertainty about which sentences express datum elements (the subject of one or more datums might be mentioned dozens of time in an article), this policy heuristically orients datum production around figure captions and sentences containing figure references: datums are aligned to such sentences, using their source field, and the machine is instructed to load its registers and produce datums as close as possible to the sentences identified in this way. For example, if a datum having subject a and treatment b is linked to sentence s_i, and b is mentioned in s_i, but the nearest mention of a is in s_{i-1}, the oracle instructs the machine to load its subject register at the a mention in s_{i-1}, and its treatment, assay, and change registers at the b mention in s_i, followed by a `produceDatums` instruction (and typically some combination of `delete` instructions).

In our current implementation, we train a multi-class perceptron model to perform ranking, updating it whenever it ranks an inappropriate instruction highest. The mistake-driven nature of this training regime enables us to accommodate a subtlety of the problem: there are often several good instructions in any given state, and we cannot know that the instruction preferred by the oracle is truly optimal. To respond to this reality, the oracle provides a second service—assessment of instructions preferred by the model. If such an instruction is deemed adequate—if it does not ultimately prevent the register machine from producing upcoming datums—the model's preferred instruction is deemed correct, and no update is performed.

Any feature of the machine's state, including the contents of its registers, datums produced so far, recently executed instructions, and, most importantly, the language at and around cursors, may be encoded to train the model. Table 2 lists the features implemented to date, which should be self-explanatory, except for the "Pattern" features. To implement these, we separately induce a set of patterns over dependency parses to detect expressions

Type	Feature	Description
Cursor	`atPosition(curs, pos)`	True if the cursor `curs` (body or caption) is at the indicated `pos` in its section (beginning, internal, end)
Register	`populated(reg)`	True if the indicated `reg` (subject, treatment, assay, or change) is populated.
	`cregContains(reg, val)`	True if a closed-class register `reg` (assay or change) contains a particular value `val` legal for that type (e.g. the assay register contains "phos").
	`oregContains(curs, reg)`	True if the open-class register `reg` contains the entity under the cursor `curs`.
	`allPopulated`	True if all four registers are populated.
Lexical	`sentContains(curs, word)`	True if the sentence under `curs` contains `word`.
	`wordAtOffset(curs, offs, word)`	True if `word` is observed at offset `offs`, ranging over $[-2, +2]$, from `curs`.
Pattern	`activeAtSent(pat, curs)`	True if the detection pattern `pat` matches the sentence under `curs`.
	`activeAtEnt(pat, curs)`	True if the pattern `pat` matches the entity under `curs`.
Other	`producedDatums`	True immediately after a produceDatums instruction has been executed.
	`bias`	Always true.

Table 2: Features used in experiments with the register machine.

that tend to signal the presence of an assay subject or treatment (Freitag and Niekrasz, 2016). For each such pattern, we define two features, which are true if the corresponding pattern matches anywhere in a cursor sentence or at a cursor entity, respectively. In addition to the features listed in the table, we automatically generate a large number of conjunctive features from feature pairs, returning true when both the constituent features are true.

5 Evaluation

We constructed our experimental data from the set of datums in the Pathway Logic database, along with the 2,394 papers to which they refer. Because most of these papers are available only as PDF,[3] we converted them to plain text and heuristically identified paper sections, converting each to a sequence of sentences. This data was then annotated by machine to identify mentions of protein entities (heuristically mapped to Uniprot identifies) and figure references. The latter were used to align datums heuristically to sentences.

As noted previously, the Pathway Logic data

	All	Phos
Database	17,444	4,864
Experimental corpus	6,554	3,152
Visible	5,981	2,989
Fully visible	2,336	1,418

Table 3: Visibility of datums (of any type vs. those representing phosphorylation assays).

comes with no guarantee that the datums are actually described in the text of an article.[4] Moreover, failures in entity recognition or resolution further reduce what our models have the potential to "see" in the text. We therefore limit our attention to "visible" datums, those datums for which we recognize either the subject or treatment entity somewhere in the paper to which a datum is aligned. We call datums for which both entities are recognized "fully visible." Our experimental corpus consists of the 518 papers aligned to at least one visible datum.

---

[3]Much of the curated data predates the establishment of the NXML format.

[4]Nor is there a strong guarantee that all experiments described in a paper have been converted into datums. Our curator has it in her charter to do so, but we have encountered experiments for which no datum was created. We do not know how common this is.

Method	Precision	Recall	F1
Oracle	0.6935	0.5708	0.5996
Frame Gold	0.9302	0.4937	0.6017
Frame	0.2426	0.296	0.2322
Machine	0.2056	0.1877	0.165

Table 4: Macro-averaged precision, recall, and F1 in extracting simplified "phos" datums.

Table 3 provides an overview of the data we work with. For convenience in comparing our two approaches, we focus on "phos" datums exclusively, and therefore present separate totals for "phos" datums in the table. (The register machine targets all visible assay types, but we evaluate its performance only against "phos" datums.) The row labeled *Database* lists counts calculated from our snapshot of the datum database, while *Experimental corpus* considers only the subset aligned to papers in our collection of 518 articles. The rows *Visible* and *Fully visible* document the number of datums actually available for experiments. The performance numbers that follow correspond to those datums contained in the cell labeled *Visible, Phos*.

In our experiments, we randomly sampled 75% of our 518 articles (and the corresponding datums) for training, and evaluated against the remaining 25%. In training the register machine, we reserve some of the training data for validation, using F1 against this hold-out data as a stopping criterion to prevent overfitting. To be deemed correct, an extracted simplified datum must agree with a ground-truth datum on all four slots. When a ground-truth datum is partially visible, an extractor must populate the empty slot with a null in order to be awarded credit. Note that this necessarily limits the recall of frame classification, which has no way to produce a null slot.

Table 4 presents the results of our experiments. The first two rows in the table establish approximate upper bounds on performance. *Oracle* measures the performance of the policy used to generate training data for the register machine, while *Frame Gold* lists the performance of a perfect classifier of candidate protein pairs nominated using the sentence co-occurrence heuristic. Interestingly, the difference in recall between these two approaches is fairly small, indicating that although the register machine can in principle integrate evidence distributed over multiple sentences, it is difficult to do so, even for a heuristically implemented oracle.

The remaining two rows compare the two learning approaches to datum extraction described in the paper, frame classification (*Frame*) and the register machine (*Machine*). Note that the example generation procedure used in Frame leads to considerable class skew, with the set of negative example dwarfing the positive. In these experiments, we randomly sampled the negative examples to achieve a ten-to-one negative-to-positive ratio.

The results appear to suggest that the relative simplicity of frame classification more than compensates for the fact that it cannot account for a significant fraction of datums, those whose subjects and treatments are not found together in an individual sentence. We see clear evidence that accumulation of evidence spread across sentences enhances performance. In a separate experiment, in which we classified individual sentences (similar to canonical relation extraction), we saw a drop in F1 of about 2 points.

The register machine, which in principle can accommodate the "distributed" datums that the frame classifier ignores, has difficulty learning instruction sequences well enough to achieve comparable performance. Its difficulty appears to center primarily on the extracted components of datums, the subjects and treatments. If we evaluate the register machine's performance on individual slots (e.g., by scoring the set of phos subjects extracted against the set found in phos datums aligned to a paper), we observe F1s of 0.92 and 0.67 on assay and change, respectively, but only 0.42 and 0.38 on subject and treatment. We believe that the feature set currently employed by the machine is too impoverished to perform these extractions accurately. Note that while frame classification accumulates evidence relevant to a protein pair from across an article, the register machine relies on mostly local information. This is an unnecessary limitation, which we are attempting to rectify.

6 Discussion

Our work with the frame classifier is leading the way in this regard. In preliminary work conducted after the experiments presented here, we have continued to mitigate keys drawbacks of the approach. For example, by training individual protein classifiers for "subjectness" and "treatmentness," using information distributed across an article, we observe a frame classification F1 of 0.30 in preliminary experiments. We are also working to increase the number of assay-change combinations targeted by frame classification to practical levels.

All this makes clear that the strict evaluation metric used in this paper–simultaneous agreement on four key slots with target datums–poses a stiff challenge for computer readers. These performance levels are understandable. Robust solutions for many types of *binary* relation and event extraction have yet to be reported. For example, a characteristic approach to ACE-style relation extraction reports peak F1 of about 0.55 (GuoDong et al., 2005), and recent work in comparable biomedical extraction problems yields qualitatively comparable performance–e.g., F1 of 0.53 in a pathway curation task involving primarily binary interactions having high domain overlap with the current paper (Nédellec et al., 2013). As a rule, adding slots to a target template leads to considerably lowered extraction performance under a strict matching regime. Moreover, the heuristic alignment of slot values to specific textual expressions adds further noise to the training and evaluation processes.

However, there is reason to believe that even these modest performance numbers are useful for certain applications. In separate work under the DARPA Big Mechanism program, we implemented a manual datum extractor, as part of a system that sought to confirm events and relations extracted by general-purpose bio-NLP readers by looking for corroborative experiments in the same paper. We were able to show, using hand-scored results from the program evaluation, that 80% of machine extractions corroborated in this way were correct (about 17% of all such extractions), versus a baseline accuracy of 50%. This despite the fact that we estimated the F1 of our the hand-authored system, which over-generates wildly, at less than 0.02. Thus, even a very noisy experiment extractor has value as a source of corroboration for assertions extracted without attention to pragmatic context. Possibly key to this outcome was the strict standard applied in the program evaluation, which deprecated speculation or statements of background knowledge.

Our focus on a very specific problem and data set may leave the impression that these results are of little further use. We argue that the opposite is true, that this admittedly domain-specific challenge is an instance of a type of problem that will become increasingly salient as machine reading matures. Eventually, the field must move beyond sentence-local, contextless, low-arity extraction to the full population of knowledge frames summarizing information relevant to important use cases. A key resource to this end will be "found" structured resources loosely attached to textual source material, such as the auxiliary data associated with biological publications with increasing frequency, or Wikipedia info-boxes summarizing events in newswire (Reschke et al., 2014). The field requires methods that exploit such resources for the interpretation of key facts in text.

7 Conclusion

The problem introduced in this paper—that of extracting faithful representations of experiments described in the biological literature—has two features that distinguish it from much of the work on biomedical NLP: (1) It is closely aligned to the needs of computational biology, stemming from research independent from and uninformed by NLP. And (2) it cannot be adequately addressed by models that target the information found in individual sentences in isolation. These two features make for a problem of considerable depth and importance, both for biology and NLP. While it is clear that we have not solved this problem with the approaches documented here, we have sketched two potential solutions and illuminated some of the specific challenges that remain.

Acknowledgments

This project was supported by the U.S. Army Research Office. The content of this paper does not necessarily reflect the position or policy of the U.S. Government. No official endorsement should be inferred.

References

Gully APC Burns, Pradeep Dasigi, Anita de Waard, and Eduard H Hovy. 2016. Automated detection of discourse segment and experimental types from the text of cancer pathway results sections. *Database* 2016.

Nancy Chinchor, David D Lewis, and Lynette Hirschman. 1993. Evaluating message understanding systems: an analysis of the third message understanding conference (muc-3). *Computational linguistics* 19(3):409–449.

Mark Craven and Johan Kumlien. 1999. Constructing biological knowledge bases by extracting information from text sources. In *ISMB*. volume 1999, pages 77–86.

P. Dasigi, G. A. P. C. Burns, E. Hovy, and A. de Waard. 2017. Experiment Segmentation in Scientific Discourse as Clause-level Structured Prediction using Recurrent Neural Networks. *ArXiv e-prints* .

Hal Daume III, John Langford, and Daniel Marcu. 2009. Search-based structured prediction. *Machine learning* 75(3):297–325.

Steven Eker, Merrill Knapp, Keith Laderoute, Patrick Lincoln, and Carolyn Talcott. 2004. Pathway logic: Executable models of biological networks. *Electronic Notes in Theoretical Computer Science* 71:144–161.

Dayne Freitag and John Niekrasz. 2016. Feature derivation for exploitation of distant annotation via pattern induction against dependency parses. In *Proceedings of the 15th Workshop on Biomedical Natural Language Processing*. Association for Computational Linguistics, Berlin, Germany, pages 36–45.

Ralph Grishman and Beth Sundheim. 1996. Message understanding conference-6: A brief history. In *Proceedings of the 16th conference on Computational linguistics-Volume 1*. Association for Computational Linguistics, pages 466–471.

Zhou GuoDong, Su Jian, Zhang Jie, and Zhang Min. 2005. Exploring various knowledge in relation extraction. In *Proceedings of the 43rd annual meeting on association for computational linguistics*. Association for Computational Linguistics, pages 427–434.

J.-D. Kim, Tomoko Ohta, Yuka Tateisi, and Junichi Tsujii. 2003. GENIA corpusa semantically annotated corpus for bio-textmining. *Bioinformatics* 19(suppl 1):i180–i182.

Jin-Dong Kim, Tomoko Ohta, and Jun'ichi Tsujii. 2008. Corpus annotation for mining biomedical events from literature. *BMC bioinformatics* 9(1):1.

Jin-Dong Kim, Yue Wang, Toshihisa Takagi, and Akinori Yonezawa. 2011. Overview of genia event task in bionlp shared task 2011. In *Proceedings of the BioNLP Shared Task 2011 Workshop*. Association for Computational Linguistics, Stroudsburg, PA, USA, BioNLP Shared Task '11, pages 7–15.

Claire Nédellec, Robert Bossy, Jin-Dong Kim, Jung-Jae Kim, Tomoko Ohta, Sampo Pyysalo, and Pierre Zweigenbaum. 2013. Overview of bionlp shared task 2013. In *Proceedings of the BioNLP Shared Task 2013 Workshop*. Association for Computational Linguistics Sofia, Bulgaria, pages 1–7.

Sampo Pyysalo, Filip Ginter, Juho Heimonen, Jari Bjrne, Jorma Boberg, Jouni Jrvinen, and Tapio Salakoski. 2007. BioInfer: a corpus for information extraction in the biomedical domain. *BMC bioinformatics* 8(1):50.

Kevin Reschke, Martin Jankowiak, Mihai Surdeanu, Christopher D. Manning, and Daniel Jurafsky. 2014. Event Extraction Using Distant Supervision. In *LREC*. pages 4527–4531.

Paul Thompson, Syed A. Iqbal, John McNaught, and Sophia Ananiadou. 2009. Construction of an annotated corpus to support biomedical information extraction. *BMC bioinformatics* 10(1):1.

Jun Xu, Yonghui Wu, Yaoyun Zhang, Jingqi Wang, Hee-Jin Lee, and Hua Xu. 2016. Cd-rest: a system for extracting chemical-induced disease relation in literature. *Database* 2016.

Classification based extraction of numeric values from clinical narratives

Maximilian Zubke
Hochschule Hannover
Dept. of Information and Communication
Expo Plaza 12, 30539 Hannover
`maximilian.zubke@hs-hannover.de`

Abstract

The robust extraction of numeric values from clinical narratives is a well known problem in clinical data warehouses. In this paper we describe a dynamic and domain-independent approach to deliver numerical described values from clinical narratives. In contrast to alternative systems, we neither use manual defined rules nor any kind of ontologies or nomenclatures. Instead we propose a topic-based system, that tackles the information extraction as a text classification problem. Hence we use machine learning to identify the crucial context features of a topic-specific numeric value by a given set of example sentences, so that the manual effort reduces to the selection of appropriate sample sentences. We describe context features of a certain numeric value by term frequency vectors which are generated by multiple document segmentation procedures. Due to this simultaneous segmentation approaches, there can be more than one context vector for a numeric value. In those cases, we choose the context vector with the highest classification confidence and suppress the rest.

To test our approach, we used a dataset from a german hospital containing 12 743 narrative reports about laboratory results of Leukemia patients. We used Support Vector Machines (SVM) for classification and achieved an average accuracy of 96% on a manually labeled subset of 2073 documents, using 10-fold cross validation. This is a significant improvement over an alternative rule based system.

1 Introduction

Driven by the digitalization, also hospitals have begun to process their documentation more and more in a digital manner. The resulting databases establish new opportunities for efficient analysis of patient data. However, many parts of those data are described by a free text, so that concrete information first has to be extracted from text before they become available for further analysis. This paper focuses on the extraction and correct semantic interpretation of numeric values from clinical narratives. Indeed, some numeric values like in example *E:G-Verhältnis=0,4:1* can extracted by regular expression or template filling due to unambiguous formattings or keywords. But there are also numeric values, which are difficult to process on that way. Reasons for the complexity are general number descriptions, like e.g. percentage values, or a variety of keywords for the associated, semantic information. In front of many different medical areas with different informations and formulations, we assume that machine learning can be used to simplify and improve this task.

After an overview of related work in section 2, we introduce a method to assign numeric values of a given document to their semantic meanings in section 3. In contrast to rule-based systems, we use a system that is able to learn and identify descriptive context features for certain numeric values by example sentences. We consider this task as a supervised machine learning problem and examine the feasibility to replace rule based systems by a more flexible machine learning approach. In section 5 we compare a rule based system with our approach and substantiate our recommendation to use machine learning procedures for information extraction processes.

Proceedings of the Biomedical NLP Workshop associated with RANLP 2017, pages 24–31,
Varna, Bulgaria, 8 September 2017.

2   Related Work

There are various research activities in the field of clinical text mining which can be divided into research in the field of Information Retrieval and research in the field of Information Extraction. We position our work in the field of Information Extraction. In general, Information Extraction in context of medical text mining often addresses one of the following tasks:

- Named Entity Recognition (Ruch et al., 2003)

- Negation Detection (Elkin et al., 2005)

- Temporal Information (Hripcsak et al., 2005)

- Extraction of Codes (ICD,OPS) (Baud, 2003)

We noticed that most of the related studies use *regular expressions* and some kind of terminology, dictionary or ontology. Especially, a robust mapping (Sager et al., 1994) between clinical narratives and *UMLS* (Lindberg, 1990), *SnomedCT* or a self-defined coding scheme appear to be the frequent goals of research in this field. Using annotation engines like *GATE* or *UIMA* text parts are connected to the corresponding concept of the given knowledge organization system (Liu et al., 2005).

In addition, some authors define or describe a complete natural language processing tool for clinical narratives, that integrates typical text mining operations like tokenization, POS-Tagging to enhance the process of information extraction. Besides *MedLEE* (Friedman et al., 1995), *Apache cTakes* (Savova et al., 2010) is such a software solution that combines the concepts, mentioned above.

It should be noticed, that many knowledge organization systems, like e.g. *SnomedCT*, are not directly available for german. Thus Becker and Böckmann (2016) describe an approach to extract *UMLS* concepts from german clinical notes using the german version of *UMLS* and find the corresponding *SnomedCT* concept by the previously detected *UMLS* concept.

Summarizing, we observe that mapping of documents to knowledge organization systems like *UMLS* or *SnomedCT*, supported by classical text mining operations, seems to be the most common approach for information extraction from clinical narratives. One often mentioned argument against the use of *machine learning* is the high effort to generate suitable training sets.

3   Method

Instead of executing a traditional Natural Language Processing (NLP) pipeline and process each word, e.g. by associating it with an *UMLS* concept, we are only interested on numeric values specified in text documents. Hence, we introduce a method to determine the meaning of a numeric value by the surrounding words using *machine learning* algorithms. This approach represents an alternative to the explicit definition of information extraction rules or ontology based document processing.

As illustrated in figure 1, our information extraction method consists of five steps:

1. Extraction of numeric values

2. Document segmentation by . and ;

3. Generation of description candidates for each numeric value

4. Classification of candidates

5. In case of multiple positive classified candidates: Suppression of all candidates, except the one with highest score.

Furthermore we use topic-based classifiers. Each topic, like i.e. *Blasts* have to be described by positive an negative example sentences. Based on this sentence sets the topic classifier determines, if a given documents belongs to that topic or not. The mentioned processing steps are explained in detail below. The performance of this approach can be found in section 5. Further details about our implementation are described in section 4.

3.1   Initial Extraction of numeric values

Because we aim to extract numeric values from clinical narratives, we are only interested in documents of the corpus C that contain at least one numerical value. Therefore we use *regular expressions* to detect and extract numerical intervals or single values from every document. The result of this initial filtering is a subset $C_{num} \subseteq C$. After this initial processing step each document $d_i \in D$ is defined as

$$d_i := (t, N_i) \tag{1}$$

where $t \in C_{num}$ represents the original text and N_i the set of numerical values that appears in that document.

d
numbers
phrases
readDoc
extractNumbers
extractPhrases
extractSentences
sentences
candidates
matches
bestMatch
n
generateCandidates
detectTopic
NMS
Topic Definition

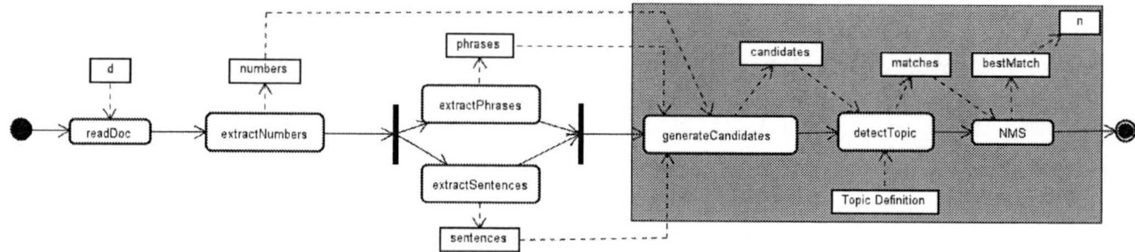

Figure 1: (1) Extraction of numeric values from document d (2) segmentation of sentences and phrases of d (3) for each n (gray area): sentence and phrases that contain n are candidates (4) topic related candidates are matches (5) Choose the match with the highest confidence

3.2 Document Segmentation

In simple clinical information systems, an unstructured text is often represented by a string. However, for advanced information extraction strings do not fit very well. Thus, the transformation of a string in a more complex data structure is the initial processing step of many text mining applications. There are several concepts to represent a document by such a complex data structure. Beside graph-based approaches (Jiang et al. (2010)), a document can also be described by bag of words or a collection of sentences.

As illustrated in Figure 2, we believe, that a numeric value is more related to certain segments like sentences or phrases and less to the whole document. Furthermore we assume, that different

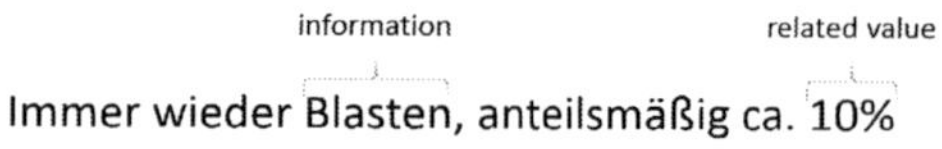

information
related value
Immer wieder Blasten, anteilsmäßig ca. 10%

Figure 2:

generated text segments could be different expressive descriptions of the contained numeric value. Due to this assumptions we describe a text d_i both as a set of sentences D_i^s and as a set of phrases D_i^p. The elements of D_i^s are produced by a common *sentence tokenizer* which splits the document into n sentences based on the dot-sign(.) without destroying point numbers or abbreviations. The elements of D_i^p are the result of the same procedure, which separates a document by semicolon instead of a dot-sign. It should be noted, that in our context the term *phrase* means a document snippet that results from the semicolon based splitting of the document. There are two motivations for this additional segmentation: First, many clinical narratives are more written like a note and less like a formal, well structured document. Therefore, it can happen, that a document transports several informations which are separated by semicolons, but do not contain any dot-signs. In those short documents, a pure dot-sign based segmentation would fail and the whole document would be considered as the related context of a certain numeric value. Second, it is possible, that an author describes a documented quantity by a dedicated sentence, but also by the beginning of the following sentence. This related part of the following sentence is usually separated by a semicolon from the rest of the sentence. An example of such a situation can be seen in figure 3.

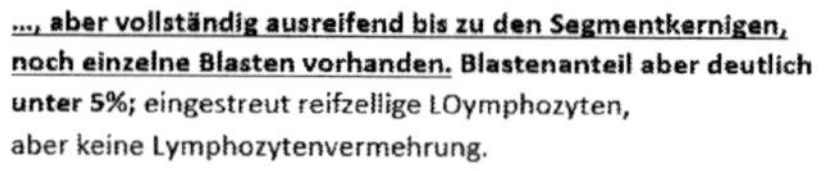

..., aber vollständig ausreifend bis zu den Segmentkernigen,
noch einzelne Blasten vorhanden. Blastenanteil aber deutlich
unter 5%; eingestreut reifzellige LOymphozyten,
aber keine Lymphozytenvermehrung.

Figure 3: Underlined: Result from pure dot-sign-based segmentation; Bold: Relevant text snippet which is delivered by semicolon based segmentation.

So finally, we have extended our definition of a document 1 to:

$$d_i := (D_i^s, D_i^p, N_i) \tag{2}$$

for all $d_i \in D$. It is possible to extend this concept by a comma based document splitting. But we omitted it due to many for our use case useless segments.

3.3 Candidate Generation

After the generation of overlapping document segments, we are only interested on segments, which are related to a numeric value n_j of d_i. Due to the use of multiple segmentation procedures, there can be more than one snippet which is directly related to n_j. We call such segments *candidates*.

In our current version, a related text segment of a numerical value n_j of document d_i can only be a sentence or phrase from the same document that contains this value, so that the *candidate set* of each $n_j \in N_i$ is defined as:

$$cand(n_j) := \{c|((c \in D_i^s) \vee (c \in D_i^p)) \wedge (n_j \in c)\} \quad (3)$$

In our implementation we keep track of relations between numerical values, sentences and phrases of d_i, so that we are able to retrieve the correct candidates even if the same numerical value appears multiple times in d_i.

3.4 Topic Learning

Usually, quantities and their numerical values appear in the same sentence or text region. It is however extremely hard to define the exact construction in which the quantity and the value appear. Consider e.g. the following sentence:

> (1) Immer wieder Blasten, anteilsmäßig ca. 10%
> *Again and again blasts, rate approx. 10 %*

The quantity *Blastenanteil* (Blast rate) is expressed in two words. The second (*Anteil*) is only present as the root of a derived adjective (*anteilsmäßig*). Patterns like this are hard to capture in rules. However, when the key concept *blasts* and a numerical value appear in the same region of the text, we can almost be sure, that the number is the value for the blast rate. To recognize such a key concept or topic, our system learns the related words by a set of sample sentences.

Our system does not have any kind of knowledge from a connected ontology or terminology base like *UMLS*. Also text mining operations like *Named Entity Recognition* or *Negation detection* are not part of our processing pipeline.

Instead our system is based on a generic concept of topic definition only. In our context a topic associated with a quantity is defined as a pair of sets containing positive and negative example sentences for numeric values of that quantity. Table 1 illustrates this idea for the amount of blasts, which is mentioned in many documents of our test dataset. Based on this two sets, we train a binary topic-classifier, which determines whether a given text segment belongs to that topic or not.

$$detect_t(c) = \begin{cases} 0 & \text{if c is not about topic} \\ (1, \kappa) & \text{if c is about topic} \end{cases} \quad (4)$$

Where κ means the confidence or score of the classification.

As already explained above, c can be a sentence or a phrase, that results from the segmentation described section 3.2

We implemented 4 by *Support Vector Machines* Boser et al. (1992). The features of all candidates are *term frequencies* of a vocabulary V, so that each candidate c is described by vector $v \in \mathbb{Z}^{|V|}$ at this point. In our experiments, V contains all words from all available clinical narratives.

We assume, that c is related to topic t, if c contains a numeric value and $detect_t(c) = 1$. The definition of κ depends on the used *machine learning* algorithm. In our experiments, κ represents the distance to the hyperplane of the SVM based classifier.

3.5 Non Maxima Suppression

The trained classifier tells, whether a document segment c belongs to a certain topic t. We assume, that the numeric value mentioned in c describes the topic-related quantity, if c belongs to t. However, the classifier could find more than one candidate relevant for the given numeric value. In such cases we select the segment with the highest confidence value and assume that the value mentioned in that segment belongs to the topic. Furthermore it is possible to identify a threshold of minimum confidence to accept a candidate as an identification of a relation between a numeric value n_j and a topic t.

4 System Description

We implemented this method as a software system, which is based on Python and SQL databases. Our system should supports simple integration into a clinical data warehouse, because many clinical narratives originate from such an information system. Furthermore, adjacent data collections could be used as features of clinical narratives or vice versa in the next version of our software.

4.1 Document representation

Before the execution of any *text mining* or *machine learning* procedure, our tool first generates a database schema like shown in Figure 4. Our in section 3.2 described segmentation concept will realized by two tables, that represent D_i^s and D_i^p. This tables are filled by scripts that implement the in section 3.2 described segmentations. Further-

Positive sample sentences	Negative sample sentences
Weiterhin Monozytoide <u>Blasten</u> (80%) bei 300 Zellen	Ca. 80-85% kleine reife Lymphozyten, einzelne mit Granula
Es findet sich eine Verdrängung der normalen Hämatopoese durch eine monomorphe <u>Blastenpopulation, die ca. 80% beträgt.</u>	Granulopoese stark linksverschoben bis zu den Promyelozyten, die ca. 35% der myeloischen Zellen ausmachen
<u>Blastenanteil</u> 2-4%	Ausreifende granulopoese mit leichter vermehrung von eosinophilen und deutlicher vermehrung von plasmazellen mit einem anteil von 5-10%, z. t. vakuolisiert; kein signifikanter <u>blastenanteil</u>

Table 1: Definition of topic "Blasts" for the quantity blast rate by positive and negative example sentences; Term-related terms are underlined. The underlining is given only for illustration here and not part of the training data.

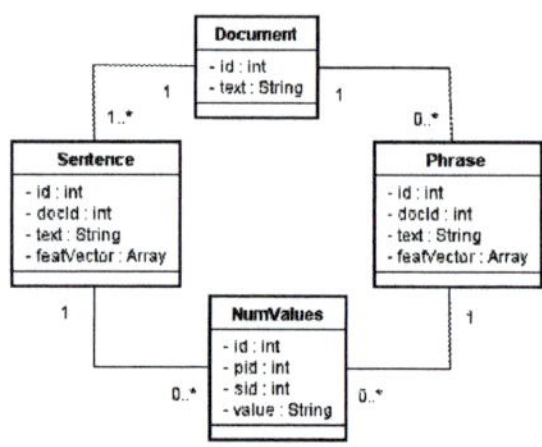

Figure 4: Documents are connected indirectly with numerical values by text segments. Each segment type is represented by a corresponding table. Currently supported segment types: Sentences and Phrases as presented in section 3.2

```
"topicName": "Blasts",
 "topicDescription": "IE of % described amount of blasts",
 "topicLanguage": "german",

    "pos":[
      "Weiterhin Monozytoide Blasten (80%) bei 300 Zellen/µL.",
      "Es findet sich eine Verdrängung der normalen Hämatopoese durch \
      eine monomorphe Blastenpopulation, die ca. 80% beträgt.",
      "Blastenanteil 2-4%."
    ],

    "neg": [
    "Ca. 80-85% kleine reife Lymphozyten, einzelne mit Granula.",
    "Vereunzelte Lymphozyten < 5%, keine pathologischen Blasten.",
    "Der Anteil an reifzelligen Lymphozyten und Plasmazellen ist deutlich \
    erhöht, stellenweise bis zu 10%."
    }
```

Figure 5: Example of our json based topic definition format.

more we store all numerical values in a dedicated table, which is filled by the procedure, we described in section 3.1. Figure 4 also illustrates, that numerical values are directly connected with sentences and phrases, but only indirectly with the documents. We chose this structure to avoid an incorrect behavior for documents, in which exactly the same numerical values appear in multiple sentences.

4.2 Topic Definition Format

We realized our in section 3.4 presented topic concept by a json based data format. Figure 5 shows an example of this technical topic description. The example sentences can be defined via an easy to use graphical user interface, that generates the appropriate json code internally. So the topics can directly defined by doctors, that do not need knowledge about technical data description techniques for this task.

A further motivation to define such a data format was the resulting flexibility, that enables the possibility to share well defined topic definitions with other internal or external organizations.

5 Evaluation & Results

We used a collection of 12 743 clinical narratives from a german hospital to evaluate our information extraction system. The narratives consist of 1 to 29 sentences, 5 sentences on average. The collection comes from electronic health records of leukemia patients. One of the main interests of the physicians is the rate of blast cells in all reports related to one patient.

At first we defined a topic by collecting positive sentences that contain a percentage description about blast cells and negative sentences that are not related to the searched topic. Example sentences for an description of the amount of blasts are:

(2) a. Blasten (80%)
   Blasts (80%)

   b. Blastenanteil 2-4%
   Blast percentage 2-4%

   c. Die Granulopoese ist linksverschoben
   mit einem Blastenanteil von $> 20\%$
   der nicht erythropoetischen Zellen
   The bone marrow is left-shifted
   with a blast proportion of $> 20\%$
   of the non erythropoietic cells.

   d. Keine Markfremden Zellen,
   Blastenanteil sicher unter 5%.
   No marrow foreign cells,
   blast percentage for sure below 5%

Then we generated a vocabulary V containing 13 400 words, based on the whole collection. A first statistic analysis shows, that the size of $|C_{num}|$ is 9 655 and only 4 162 of that documents contain known keywords about blasts and a percentage sign.

5.1 Construction of a gold standard

For the gold standard we selected a random subset of 2 073 documents, which proportion of documents is fulfilling the three conditions is the same as in the whole collection. About 75% of the documents in this selection do not contain a numerical value, or a percentage sign or a keyword related to blasts. We annotated these documents manually. Note that thus we make no difference between documents that have no information on blast rate and documents that do contain information on blast rate, but do not give a concrete value. Especially this means that we labeled all documents containing the statement *Keine Blasten* (no blasts) as documents that do not give a value for the quantity blast rate. For the remaining 435 documents, that contain keywords about blasts, a percentage sign and a numerical value, we extracted the blast percentage manually.

Our classifier is trained only on sentences containing numerical values. In our subset there are 6 805 sentences; 604 sentences contain a numerical value, 439 thereof being a blast rate, 165 not related to the amount of blasts.

5.2 Experiment setup

Each text was first split into sentences and phrases as described in section 3.2.

Next, we generated a candidate set for each numerical value that appears in the given document. As described in section 3.3, the term *candidate* means a sentence or a phrase that contains the numeric value. We processed all documents on that way.

Then we conducted two experiments: In the first experiment we examined the classification of single sentences. Beside two baselines that are described in the next section, we used a SVM based topic classifier (see section 3.4), which decides for each of the sentences, whether it is relevant for the quantity blast rate. Now we can evaluate how many sentences are classified correctly.

In the second experiment we compared methods for extracting numerical values from whole documents. We evaluated our approach in two configurations: *SVM (Sentences)* represents a variant where all elements of the candidate sets are sentences and *SVM (Sentences & Phrases)* represents the same approach using multiple text segments.

For both experiments, we consider a text as correctly processed when either (1) the correct blast rate is extracted from the text or (2) it is correctly detected that no blast rate is specified.

Our manual labeling has extracted values for each text and each sentence, obtained by splitting texts on full stops. When we make additional segments by splitting on semicolons, we can apply the classifier (trained on whole sentences) to this segments as well. However, we cannot compare the results with the manually labeled ones. On the document level, however, we can compare with the manually labeled documents.

We used ten-fold cross validation for all experiments.

5.3 Baselines

We used three baselines. Since most documents are not relevant for the quantity blast rate, we can classify most documents correctly with the majority classifier, that assumes that all documents are irrelevant.

The second baseline assumes that every percentage value is a blast rate. On the sentence level this baseline thus treats all sentences with a number and percentage sign as relevant for the blast rate and all others as irrelevant. At the document level this baseline assumes the first percentage mentioned to be the blast rate. We will refer to this baseline as the %-based approach.

As a third baseline we used an extraction method that is purely based on complex regular expressions. Motivated by the remarkable performance of the percent-based approach, a group of students developed a regular expressions based approach. Therefore they analyzed the data set and define some keywords manually. Combined with the detection of percentage values, they implemented a procedure to extract the searched informations by pattern recognition. Note that this approach processes only whole documents, which is why we could not compare this baseline with alternative approaches on sentence level described by table 2.

6 Results

Table 2 shows the result of the evaluation at sentence level We clearly observe, that the classifier treats almost all sentences correctly. With respect to precision and recall it is of course easy to beat the majority baseline, but the SVM also has an higher accuracy.

Given the good results of the %-based approach we can conclude that indeed most numerical values are related to blast rates. However, there are a number of other numerical values. Apparently, the SVM effectively distinguishes the blast rates from other numerical values.

Table 3 shows the results of the complete method on the document level. At the document level we see again very high scores. We could observe, that the additional semicolon based segmentation indeed excludes a number of mistakes. (e.g. the third negative example from Table 2) The lower precision in comparison to the pure sentence-based configuration implies, that the semicolon based approach produces a few segments which are hard to classify by the current version of our topic classifier. But *SVM(Sentences & Phrases)* also extracts significant more numeric values than *SVM(Sentences)*. As documented in table 3, the regular expression based integration of keywords improves the performance of the %-based information extraction strategy. Apparently, the rules a very precise and do almost never consider a percentage as a blast rate if that is not the case. Thus this method has the highest precision of all tested methods. However, the recall is much lower than that of the classifier based approach.

7 Conclusions and Future Work

In this paper we presented a first version of our information extraction system for medical documentations, which identifies the meaning of a numeric value by the surrounding words.

The integral difference to many similar applications is, that we had no explicit described knowledge about the content of out dataset. Instead we used *machine learning* to learn important keywords by sample sentences.

With term frequency vectors, we used a very simple kind of feature, which already works very well. In the future we want to examine, which alternative features could improve our system.

Our approach yields remarkable results. However, there are situations, that can not processed correctly by our system. We expect, that numerical values are always described by numbers. However, it is possible, that numbers are described by a words instead of number (i.e 'five' instead 5). We also observed, that especially the number zero is often replaced by a negation (i.e. 'no blasts' instead of '0% blasts'). Hence we will integrate a preprocessing step that converts textual definitions of numbers in real numbers. It should be noted, that this task is a non-trivial task, because also a quantitative value can correspond with several, very different formulation, which can be considered as an classification problem, very similar to our topic detection problem, described in section 4. Furthermore, words like 'significant' complicate or prevent a mapping to an equivalent numerical description of the information.

In general, we believe that *machine learning* could be much more efficient than rule-based concepts. Every rule engine needs someone who defines suitable rules, whereas our approach only needs sample sentences which are always available. Furthermore table 3 shows, that the *machine learning* approach is more adjustable than the more strict rule-based approach.

Acknowledgements

We would like to thank our colleagues from the Hannover Medical School to suggest the problem of extracting numerical values and making available the pseudonymized texts. Further Acknowledgements go to our students, that implemented parts of the system, along with a user interface for practical usage of the system in the Hannover Medical School hospital.

Method	Recall	Precision	Accuracy
SVM	0.987 (0.005)	0.950 (0.003)	0.996 (0)
Majority	0.0 (0)	0.0 (0)	0.935 (0)
%-based	0.893 (0)	0.727 (0)	0.971 (0)

Table 2: Results of the extraction of the percentage of blasts evaluated on **sentence** level. Results are averages of 10-fold cross-validation. Standard deviations are given in parentheses.

Method	Recall	Precision	Accuracy
SVM (Sentences & Phrases)	0.921 (0.049)	0.911 (0.044)	0.965 (0.017)
SVM (Sentences)	0.834 (0.069)	0.953 (0.037)	0.957 (0.017)
RegExp based	0.517 (0.053)	0.983 (0.021)	0.897 (0.019)
%-based	0.461 (0.082)	0.629 (0.081)	0.897 (0.023)
Majority	0.0 (0)	0.0 (0)	0.79 (0.034)

Table 3: Results of the extraction of the percentage of blasts evaluated on **document** level. Results are averages of 10-fold cross-validation. Standard deviations are given in parentheses.

References

R Baud. 2003. A natural language based search engine for icd10 diagnosis encoding. *Medicinski arhiv* 58(1 Suppl 2):79–80.

M Becker and B Böckmann. 2016. Extraction of umls® concepts using apache ctakes™ for german language. *Studies in health technology and informatics* 223:71–76.

Bernhard E Boser, Isabelle M Guyon, and Vladimir N Vapnik. 1992. A training algorithm for optimal margin classifiers. In *Proceedings of the fifth annual workshop on Computational learning theory*. ACM, pages 144–152.

Peter L Elkin, Steven H Brown, Brent A Bauer, Casey S Husser, William Carruth, Larry R Bergstrom, and Dietlind L Wahner-Roedler. 2005. A controlled trial of automated classification of negation from clinical notes. *BMC medical informatics and decision making* 5(1):13.

Carol Friedman, Stephen B Johnson, Bruce Forman, and Justin Starren. 1995. Architectural requirements for a multipurpose natural language processor in the clinical environment. In *Proceedings of the Annual Symposium on Computer Application in Medical Care*. American Medical Informatics Association, page 347.

George Hripcsak, Li Zhou, Simon Parsons, Amar K Das, and Stephen B Johnson. 2005. Modeling electronic discharge summaries as a simple temporal constraint satisfaction problem. *Journal of the American Medical Informatics Association* 12(1):55–63.

Chuntao Jiang, Frans Coenen, Robert Sanderson, and Michele Zito. 2010. Text classification using graph mining-based feature extraction. *Knowledge-Based Systems* 23(4):302–308.

C Lindberg. 1990. The unified medical language system (umls) of the national library of medicine. *Journal (American Medical Record Association)* 61(5):40–42.

Kaihong Liu, Kevin J Mitchell, Wendy W Chapman, and Rebecca S Crowley. 2005. Automating tissue bank annotation from pathology reports–comparison to a gold standard expert annotation set. In *AMIA Annual Symposium Proceedings*. American Medical Informatics Association, volume 2005, page 460.

Patrick Ruch, Robert Baud, and Antoine Geissbühler. 2003. Using lexical disambiguation and named-entity recognition to improve spelling correction in the electronic patient record. *Artificial intelligence in medicine* 29(1):169–184.

Naomi Sager, Margaret Lyman, Ngo Thanh Nhan, and Leo J Tick. 1994. Automatic encoding into snomed iii: a preliminary investigation. In *Proceedings of the Annual Symposium on Computer Application in Medical Care*. American Medical Informatics Association, page 230.

Guergana K Savova, James J Masanz, Philip V Ogren, Jiaping Zheng, Sunghwan Sohn, Karin C Kipper-Schuler, and Christopher G Chute. 2010. Mayo clinical text analysis and knowledge extraction system (ctakes): architecture, component evaluation and applications. *Journal of the American Medical Informatics Association* 17(5):507–513.

Understanding of unknown medical words

Natalia Grabar
CNRS UMR 8163 STL,
Université Lille 3,
59653 Villeneuve d'Ascq, France
`natalia.grabar@univ-lille3.fr`

Thierry Hamon
LIMSI, CNRS, Université Paris-Saclay,
Orsay, France
Université Paris 13, Sorbonne Paris Cité,
Villetaneuse, France
`hamon@limsi.fr`

Abstract

We assume that unknown words with internal structure (affixed words or compounds) can provide speakers with linguistic cues as for their meaning, and thus help their decoding and understanding. To verify this hypothesis, we propose to work with a set of French medical words. These words are annotated by five annotators. Then, two kinds of analysis are performed: analysis of the evolution of understandable and non-understandable words (globally and according to some suffixes) and analysis of clusters created with unsupervised algorithms on basis of linguistic and extra-linguistic features of the studied words. Our results suggest that, according to linguistic sensitivity of annotators, technical words can be decoded and become understandable. As for the clusters, some of them distinguish between understandable and non-understandable words. Resources built in this work will be made freely available for the research purposes.

1 Introduction

Often, people face unknown words, be they neologisms (like in *Some of the best effects in my garden have been the result of serendipity.*) or technical words from specialized areas (like in *Jacques Chirac's historic corruption trial, due to start on Monday is on the verge of collapse, after doctors diagnosed him with "anosognosia"*). In both cases, their semantics may be opaque and their understanding not obvious.

Several linguistic operations are available for enriching the lexicon, such as affixation, compounding and borrowings (Guilbert, 1971). We are particularly interested in words with internal structure, like *anosognosia*, because we assume that linguistic regularities (components, affixes, and rules that form their structure) can help speakers in deducing their structure and semantics. Our hypothesis is that if regularities can be observed at the level of linguistic features, they can also be deduced and managed by speakers. Indeed, linguistic understanding is related to factors like:

- knowledge and recognition of components of complex words: how to segment words, like *anosognosia*, in components;

- morphological patterns and relations between components: how to organize the components and to construct the word semantics (Iacobini, 2003; Amiot and Dal, 2008).

To verify our hypothesis, we propose to work with a set of French medical words. These words are considered out of context for several reasons:

1. when new words appear, they have little and poor contexts, which cannot usually help their understanding;

2. similarly, in specialized areas, the contexts, except some definitional contexts, often bring little help for the understanding of terms;

3. working with words out of context permits to process a bigger set of words and to make observations with larger linguistic material;

4. from another point of view, analysis of words in context corresponds to their perception *in extension* relying on external clues, while analysis of words out of context corresponds to their perception *in intension* relying on clues and features internal to these words.

For these reasons, we assume that internal structure of unknown words can help their understanding. According to our hypothesis, affixed words

Proceedings of the Biomedical NLP Workshop associated with RANLP 2017, pages 32–41,
Varna, Bulgaria, 8 September 2017.

and compounds, which are given internal structure, can provide the required linguistic clues. Hence, the speakers may linguistically analyze unknown words thanks to the exploitation of their structure that they are able to detect.

Our interest for medical words is motivated by an increasing presence of medical notions in our daily life, while medicine still keeps a lot of mysteries unknown to lay persons because medical knowledge is typically encoded with technical and very specialized terms.

In what follows, we present some existing works (section 2), the data which we propose to process (section 3), and the experiments we propose to exploit (sections 4 to 6). We conclude with some orientations for future work (section 7).

2   Existing work

We concentrate on work related to text difficulty and understanding. Work on processing of words unknown in dictionaries by automatic applications, although well studied, is not presented.

NLP provides a great variety of work and approaches dedicated to understanding and readability of words and texts. The goal of readability is to define whether texts are accessible for readers or not. Readability measures are typically used for evaluation of document complexity. Classical readability measures exploit information on number of characters and syllables of words (Flesch, 1948; Gunning, 1973), while computational measures can involve vectorial models and different features, among which combination of classical measures with terminologies (Kokkinakis and Toporowska Gronostaj, 2006); n-grams of characters (Poprat et al., 2006); stylistic (Grabar et al., 2007) or discursive (Goeuriot et al., 2007) features; lexicon (Miller et al., 2007); morphological information (Chmielik and Grabar, 2011); and combination of various features (Wang, 2006; Zeng-Treiler et al., 2007; Leroy et al., 2008; François and Fairon, 2013; Gala et al., 2013).

In linguistics and psycholinguistics, the question on understanding of lexicon may focus on:

- Knowledge of components of complex words and their decomposition. The purpose is to study how complex words (affixed or compounds) are processed and recorded. Several factors may facilitate reading and production of complex words: when these compounds contain hyphens (Bertram et al., 2011) or spaces (Frisson et al., 2008); when they are presented with other morphologically related words (Lüttmann et al., 2011); and when primes (Bozic et al., 2007; Beyersmann et al., 2012), pictures (Dohmes et al., 2004; Koester and Schiller, 2011) or favorable contexts (Cain et al., 2009) are used;

- Order of components and variety of morphological patterns. Position of components (head or modifier) proved to be important for processing of complex words (Libben et al., 2003; Holle et al., 2010; Feldman and Soltano, 1999). The notions of semantic transparency and of *morphological headedness* have been isolated (Jarema et al., 1999; Libben et al., 2003);

- Word length and types of affixes (Meinzer et al., 2009);

- Frequency of bases and components (Feldman et al., 2004).

Our hypothesis on emerging of linguistic rules involved in word formation has also been addressed in psycholinguistics, and it has to face two other hypothesis on acquisition in context and on providing explicit information on semantics of components (Baumann et al., 2003; Kuo and Anderson, 2006; McCutchen et al., 2014). Currently, the importance of morphological structure for word processing seems to be accepted by psycholinguists (Bowers and Kirby, 2010), which supports our hypothesis. Yet, in our work, for verifying this hypothesis, we exploit NLP methods and NLP-generated features. Hence, we can work with large linguistic data and exploit quantitative and unsupervised methods.

3   Exploited data

The data processed are obtained from medical terminology Snomed International (Côté, 1996) in French, which purpose is to describe the medical area. This terminology contains 151,104 terms structured in eleven semantic axes (e.g. disorders and abnormalities, medical procedures, chemicals, leaving organisms, anatomy). We keep terms from five axes (disorders, abnormalities, medical procedures, functions and anatomy), which we consider to be central and frequent. Hence, we do not wish to concentrate on very specialized terms and

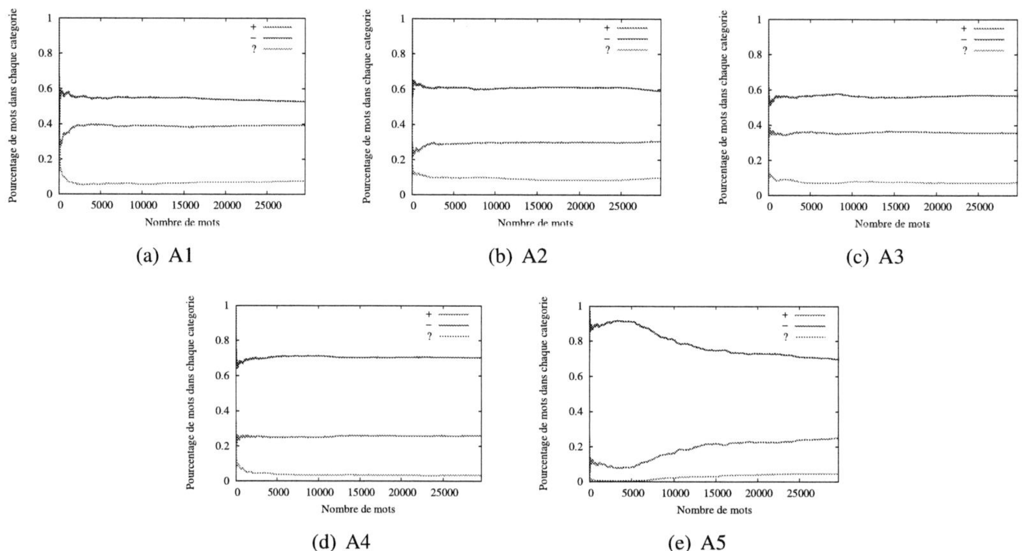

Figure 1: Global evolution of percentage of words per caterogy.

words, like chemicals or leaving organisms. Nevertheless, such words can be part of terms studied here. The selected terms (104,649) are segmented in words to obtain 29,641 unique words, which are our working material. This set contains compounds (*abdominoplastie (abdominoplasty), dermabrasion (dermabrasion)*), constructed (*cardiaque (cardiac), lipoïde (lipoid)*) and simple (*fragment*) words, as well as abbreviations (*AD-Pase, ECoG, Fya*) and borrowings (*stripping, Conidiobolus, stent, blind*).

These terms are annotated by five French native speakers, aged from 25 to 60, without medical training and with different social and professional status. Each annotator received a set with randomly ordered 29,641 words. According to the guidelines, the annotators should not use additional information (dictionaries, encyclopedia, etc.), should not change annotations done previously, should manage their time and efforts, and assign each word in one of the three categories: (1) *I can understand*, containing known words; (2) *I am not sure*, containing hesitations; (3) *I cannot understand*, containing unknown words. We assume that our annotators represent moderate readability level (Schonlau et al., 2011), i.e. the annotators have a general language proficiency but no specific knowledge in medical domain, and that we will be able to generalize our observations on the same population. Besides, we assume that

these annotations will allow to observe the progression in the understanding of technical words.

Manual annotation required from 3 weeks up to 3 months. The inter-annotator agreement (Cohen, 1960) is over 0.730. Manual annotation allows to distinguish several types of words which are difficult to understand: (1) abbreviations (e.g. , *OG, VG, PAPS, j, bat, cp*); (2) proper names (e.g. , *Gougerot, Sjögren, Bentall, Glasgow, Babinski, Barthel, Cockcroft*), which are often part of terms meaning disorders and procedures; (3) medications; (4) several medical terms meaning disorders, exams and procedures. These are mainly compounds (e.g. *antihémophile (anti haemophilus), sclérodermie (sclerodermia), hydrolase (hydrolasis), tympanectomie (tympanectomia), synesthésie (synesthesia)*); (5) borrowings; (6) words related to human anatomy (e.g. *cloacal (cloacal), nasopharyngé (nasopharyngal), mitral (mitral), diaphragmatique (diaphragmatic), inguinal (inguinal), érythème (erythema), maxillo-facial (maxillo-facial), mésentérique (mesenteric), mésentère (mesentry)*).

4 Experiments

We propose two experiments:

1. Study of understanding progression of words globally and according to some components (section 5);

2. Unsupervised classification of words, analysis of clusters and their comparison with manual annotations (section 6).

5 Progression in word understanding

Progression of word understanding corresponds to the rate of understandable and non-understandable words at a given moment t for a given annotator. This permits to observe whether the annotators can become familiar with some components or morphological rules, and improve their understanding of words while the annotation is going on. This analysis is done on the whole set of words and on words with some components.

Figure 1 indicates the evolution of the three categories of words. The line corresponding to *I cannot understand* is in the upper part of the graphs, while the line *I can understand* is in the lower part. The category *I am not sure* is always at the bottom. We can distinguish the following tendencies:

- Annotators A2, A1 and especially A5 show the tendency to decrease the proportion of unknown words. We assume that they are becoming more familiar with some components and bases, and that they can better manage medical lexicon;

- Annotators A1, and in a lesser way A2 and A4, show the tendency to decrease the number of hesitation (category 2). Indeed, the proportion of these words decreases, while the proportion of words felt as known (category 1) increases. Later, the number of known words seems not to increase, except for A5. Besides, this learning effect is especially observable with the top 2,000 words and it mainly affects the transition of hesitation words to known words;

- For annotators A3 and A4, after a small increase of proportion of unknown words, this proportion remains stable. We assume that the annotation process of a large lexicon did not allow to gain in understanding of components of the processed technical words.

Figures 2 and 3 show the evolution of understanding of words ending with *-ite (-itis)* (meaning *inflammation*) and *-tomie (-tomy)* (meaning *removal*), respectively. We can see that A5 has difficulty to understand these words: the percentage of unknown words is increasing, while on the whole set of words (figure 1(e)) this annotator shows the opposite tendency, with the percentage of unknown words decreasing. Annotators A2 and A4 also have understanding difficulties with these words. Figures of other annotators suggest that they make progress in decoding and understanding of words in *-ite* and *-tomie*. They first show an improvement in understanding of these words, and later there is another small progression. On the basis of these observations, we can see that, according to types of words, to their linguistic features and to the sensitivity of annotators, it it possible to make progressive improvement in understanding of technical lexicon which *a priori* is unknown by speakers. As already noticed, we assume that linguistic regularities play an important role in improving of the understanding of new lexicon. We propose to observe now if such regularities can also be detected by unsupervised clustering algorithms.

6 Unsupervised classification of words

Unsupervised classification is performed with several algorithms implemented in Weka: SOM (Kohonen, 1989), Canopy (McCallum et al., 2000), Cobweb (Fisher, 1987), EM (Dempster et al., 1977), SimpleKMeans (Witten and Frank, 2005). Excepting SimpleKMeans and EM, it is not necessary to indicate the expected number of clusters. Each word is described with 23 linguistic and extra-linguistic features, which can be grouped in 8 classes (an excerpt is provided in Table 1):

- *POS-tags.* POS-tags and lemmas are computed by TreeTagger (Schmid, 1994) and then checked by Flemm (Namer, 2000). POS-tags are assigned to words within the context of their terms. If a given word receives more than one tag, the most frequent is kept as feature. Among the main tags we find for instance nouns, adjectives, proper names, verbs and abbreviations;

- *Presence of words in reference lexica.* We exploit two French reference lexica: TLFi[1] and *lexique.org*[2]. TLFi is a dictionary of the French language covering XIX and XX centuries, and contains almost 100,000 entries.

---

[1] http://www.atilf.fr/
[2] http://www.lexique.org/

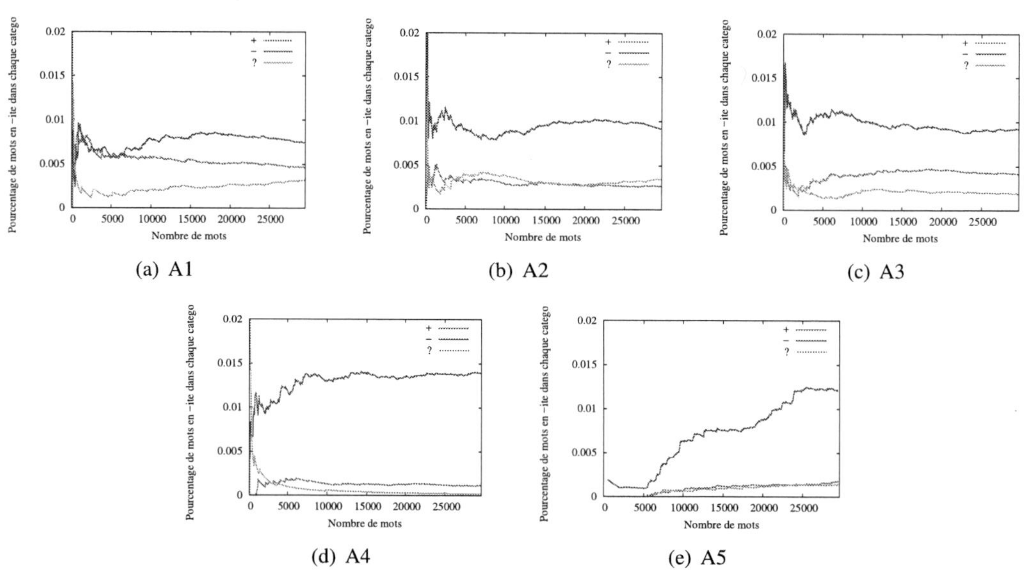

Figure 2: Evolution of percentage of words ending with *-ite* in each category.

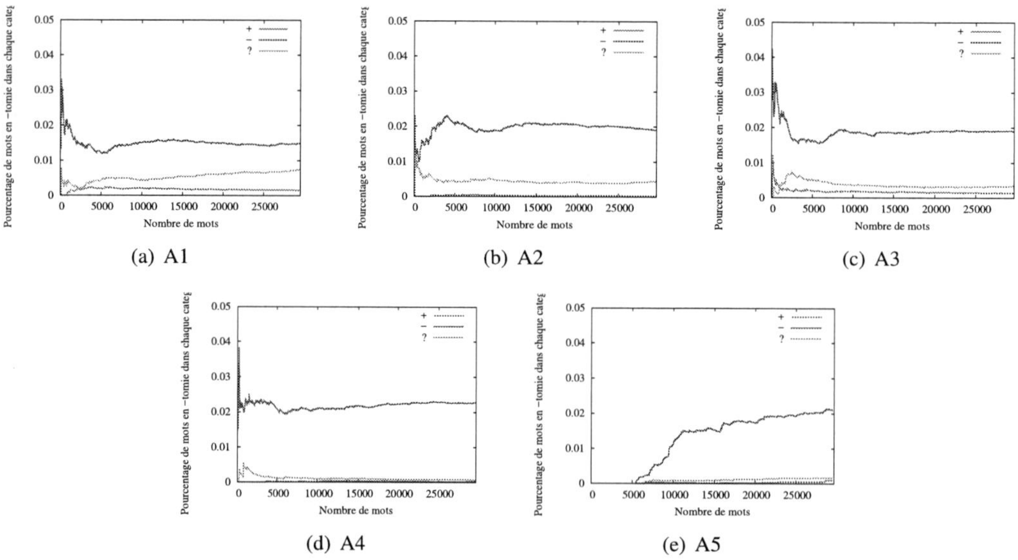

Figure 3: Evolution of percentage of words ending with *-tomie* in each category.

lemma	POS	l_1	l_2	f_g	f_t	nb_a	nb_s	initial	final	nb_c	nb_v
alarme	N	+	+	73400000	6	1	2	ala,alar,alarm	rme,arme,larme	3	3
hépatite	N	+	+	15300000	9	3	3	hép,hépa,hépat	ite,tite,atite	4	4
angiocholite	N	-	+	74700	12	1	5	ang,angi,angio	ite,lite,olite	6	6
desmodontose	N	+	-	2050	12	1	4	des,desm,desmo	ose,tose,ntose	7	5

Table 1: Excerpt with features: *POS*-tag, presence in reference lexica (TLFI l_1 and `lexique.org` l_2), frequency in search engine f_g and terminology f_t, number of semantic axes nb_a, number of syllables nb_s, initial and final substrings (*initial, final*), number of consonants nb_c, number of vowels nb_v.

lexique.org has been created for psycholinguistic experiments. It contains over 135,000 entries, including inflectional forms. It contains almost 35,000 lemmas. We assume that words that are part of these lexica may be easier to understand;

- *Frequency of words through a non specialized search engine.* For each word, we query the Google search engine in order to know its frequency attested on the web. We assume that words with higher frequency may be easier to understand;

- *Frequency of words in medical terminology.* For the same reason as above, we compute the frequency of words in the medical terminology Snomed International;

- *Number and types of semantic categories associated to words.* We also exploit the information on semantic axes of Snomed International and assume that words which occur in several axes are more central;

- *Length of words in number of characters and syllables.* For each word, we compute the number of its characters and syllables, because we think that longer words may be more difficult to understand;

- *Number of bases and affixes.* Each lemma is analyzed by the morphological analyzer Dérif (Namer and Zweigenbaum, 2004), adapted to the treatment of medical words. It performs decomposition of lemmas into bases and affixes, and provides semantic explanation of the analyzed lexemes. We exploit morphological decomposition, which permits to compute the number of affixes and bases. Here again we focus on complexity of the internal structure of words;

- *Initial and final substrings.* We compute the initial and final substrings of different length, from three to five characters. This allows to isolate some components and possibly the morphological head of words;

- *Number and percentage of consonants, vowels and other characters.* We compute the number and the percentage of consonants, vowels and other characters (*i.e.* hyphen, apostrophe, comas).

We perform experiments with three featuresets:

- E_c: the whole set with 23 features,

- E_r: set with features reduced to linguistic properties of words, such as POS-tag, number of syllables, initial and final substrings, which permits to take into account observations from psycholinguistics (Jarema et al., 1999; Libben et al., 2003; Meinzer et al., 2009),

- E_f: set with linguistic features and frequency collected with the search engine, which permits to consider other psycholinguistic observations (Feldman et al., 2004).

With `SimpleKMeans` and `EM`, we perform two series of experiments, in which the number of clusters is set to 1,000 and 2,000 (for almost 30,000 individuals to cluster). We expect to find linguistic regularities of words in clusters, according to the features exploited. More specifically, we want to observe whether the content of clusters is related to the understanding of words.

Features	SOM	Canopy	Cobweb
E_c: Full set (23)	5	62	33853
E_r: Reduced set (8)	4	28	12577
E_f: E_r and frequency (9)	4	27	9861

Table 2: Generated clusters

In Table 2, we indicate the number of clusters obtained with various sets of features: SOM generates very few clusters, which are big and heterogeneous. For instance, with E_f, clusters contain up to 13,088, 4,840, 7,023 and 4,690 individuals; Cobweb generates a lot of clusters among which several singletons. For instance, with E_f, we obtain 9,374 clusters out of which 9,861 are singletons; EM and `SimpleKMeans` generate the required number of clusters, 1,000 and 2,000; Canopy generates between 30 and 60 clusters, according to the features used. We propose to work with clusters obtained with Canopy because it generates reasonnable number of clusters, which number and contents are motivated by features.

With features from sets E_r and E_f, cluster creation is mainly motivated by initial substrings (not always equal to 3 to 5 first or final characters) and in a lesser way by their POS-tags and frequencies. For instance, we can obtain clusters with words beginning by p or a, or clusters grouping *phosphats* or enzymes ending with *-ase*. In this last case, clusters with chemicals become interesting for our purpose, although globally the clusters generated on basis of features from sets E_r and E_f show little interest. We propose to work with clusters obtained with the E_c featureset.

With `Canopy`, the size of clusters varies between 1 and 2,823 individuals. Several clusters are dedicated to two main annotation categories. Hence, 30 clusters contain at least 80% of words from the category *1 (I can understand)*, while 6 clusters contain at least 80% of words from the category *3 (I cannot understand)*. Among the clusters with understandable words, we can find clusters with:

- numerals (*mil (thousand), quinzième (fifteen)*), verbs (*allaite (breast-feed), étend (expand)*), and adverbs (*massivement (massively), probablement (probably)*) grouped according to their POS-tags and sometimes to their final substrings;

- grammatical words (*du (of), aucun (any), les (the)*) grouped on basis of length and POS-tags;

- common adjectives (*rudimentaire (rudimentary), prolongé (extended), perméable (permeable), hystérique (hysterical), inadéquat (inadequate), traumatique (traumatic), militaire (military)*) grouped according to their POS-tags and frequency;

- participial adjectives (*inapproprié (inappropriate), stratifié (stratified), relié (related), modifié (modified), localisé (localised), précisé (precise), quadruplé (quadrupled)*) grouped according to their POS-tags, frequencies and final substrings;

- specialized but frequent adjectives (*rotulien (patellar), spasmodique (spasmodic), putréfié (putrefactive), redondant (redundant), tremblant (trembling), vénal (venal), synchrone (synchronous), sensoriel (sensory)*), also grouped according to their POS-tags and frequencies;

- specialized frequent nouns (*dentiste (dentist), brosse (brush), altitude (altitude), glucose (glucose), fourrure (fur), ankylose (ankylosis), aversion (aversion), carcinome (carci-*

noma)) grouped according to their POS-tags and frequencies.

Among the clusters with non-understandable words, we can find:

- chemicals (*dihydroxyisovalérate, héparosane-N-sulfate-glucuronate, désoxythymidine-monophosphate, diméthylallyltransférase*) grouped according to their POS-tags, types of characters they contain and their frequency;

- borrowings (*punctum, Saprolegnia, pigmentosum, framboesia, equuli, rubidium, dissimilis, frutescens, léontiasis, materia, mégarectum, diminutus, ghost, immitis, folliclis, musculi*) grouped according to their POS-tags, final substrings and frequency;

- proper names grouped according to their POS-tags.

Within clusters with over 80% of words from the category *3 (I cannot understand)*, we do not observe understanding progression of annotators. Yet, we have several mixed clusters, that contain words from the two main categories (*1 (I can understand)* and *3 (I cannot understand)*), as well as hesitations. These clusters contain for instance:

- chemicals and food (*créatinine (creatinine), antitussif (antitussive), céphalosporine (cephalosporine), aubergine (eggplant), carotte (carrot), antidépresseur (antidepressant), dioxyde (dioxide)*) grouped according to their final substrings, semantic axes and frequency;

- organism functions, disorders and medical procedures (*paraparésie (paraparesis), névralgie (neuralgia), extrasystole (extrasystole), myéloblaste (myeloblast), syncope (syncope), psychose (psychosis), spasticité (spasticity)*) grouped according to their frequency, final substrings and POS-tags;

- more specialized adjectives related to anatomy and disorders (*périprostatique (periprostatic), sous-tentoriel (tensor), condylienne (condylar), fibrosante (fibrotic), nécrosant (necrosis)*) grouped according to their POS-tags and frequency.

Evolution of understanding is observable mainly within this last set of clusters. For instance, a typical example is the cluster containing medical procedures ending in *-tomie*, which words become less frequently assigned to the category *3 (I cannot understand)* and more frequently to the categories *2 (I am not sure)* and *1 (I can understand)*.

The content of clusters and our observations suggest that, given an appropriate set of features and unsupervised algorithms, it is possible to create clusters which reflect the readability and understandability of lexicon by lay persons. Besides, within some clusters, it is possible to observe the evolution of annotators in their understanding of technical words. For instance, this effect can typically be observed with words meaning disorders and procedures. Nevertheless, with other types of words (chemicals, borrowings, proper names) no evolution is observable.

Notice that the same reference data have been used with supervised categorization algorithms. In this case, automatic algorithms can reproduce the reference categorization with F-measure over 0.80 and up to 0.90, which is higher than the inter-annotator agreement rate. Besides, in the supervised categorization task, the behaviour of features is different from what we can observe in unsupervised clusters: several individual features can reproduce the reference categories while the best results are obtained with the whole set of features.

7 Conclusion and Future work

According to our hypothesis, linguistic regularities, when they occur systematically, can help in decoding and understanding of technical words with internal structure (like compounds or derived words). To test the hypothesis, we work with French medical words. Almost 30,000 words are annotated by five annotators and assigned in one of the three categories *I can understand*, *I am not sure*, *I cannot understand*. For each annotator, the words are ordered randomly.

We then perform an analysis of the whole set of words, and of words ending with *-ite* and *-tomie*. Our results suggest that several annotators show the learning effect as the annotation is going on, which supports our hypothesis and the findings of psycholinguistic work (Lüttmann et al., 2011). This effect is observed for the whole set of words and for the two analyzed suffixes. Yet, with chemicals, borrowings and proper names, we do not observe the learning effect.

These observations have been corroborated with clusters generated using linguistic and extra-linguistic features. Several clusters are dedicated to words from either *1 (I can understand)* or *3 (I cannot understand)* categories. Besides, when clusters contain some semantically homogeneous words (disorders, procedures...) we can observe the expected learning effect. These results are very interesting and confirm our hypothesis, according to which linguistic regularities can help to decode and understand technical and unknown words. Appropriate features can also help to distinguish between understandable and non-understandable words with unsupervised methods. Correlations between social and demographic status and understanding require additional annotations. It will be studied in the future.

We have several directions for future work: (1) collect the same type of annotations, but providing semantics of some or of all components, although it will be difficult to verify whether this information is really exploited by annotators; (2) collect the same type of annotations, but permitting the annotators to use external sources of informations (dictionaries, online examples...). Since this approach requires more time and cognitive effort, smaller set of words will be used; (3) analyze the evolution of understanding of words taking into account a larger set of components; (4) validate the observations with tests for statistical significance; (5) exploit the results for training and education of non-experts in order to help them with the understanding of medical notions; (6) exploit the results for simplification of technical texts. For instance, features of words that show understanding difficulties can be used to define classes of words that should be systematically simplified.

The resources built in this work are freely available for the research purposes: `http://natalia.grabar.free.fr/resources.php#rated`.

Acknowledgments

We would like to thank the Annotators for their hard annotation work. This research has received aid from the IReSP financing partner within the 2016 general project call, Health service axis (grant GAGNAYRE-AAP16-HSR-6).

References

D Amiot and G Dal. 2008. La composition néoclassique en français et ordre des constituants. *La composition dans les langues* pages 89–113.

JF Baumann, EC Edwards, EM Boland, S Olejnik, and EJ Kame'enui. 2003. Vocabulary tricks: Effects of instruction in morphology and context on fifth-grade students' ability to derive and infer word meanings. *American Educational Research Journal* 40(2):447–494.

Raymond Bertram, Victor Kuperman, Harald R Baayen, and Jukka Hyönä. 2011. The hyphen as a segmentation cue in triconstituent compound processing: It's getting better all the time. *Scandinavian Journal of Psychology* 52(6):530–544.

Elisabeth Beyersmann, Max Coltheart, and Anne Castles. 2012. Parallel processing of whole words and morphemes in visual word recognition. *The Quarterly Journal of Experimental Psychology* 65(9):1798–1819.

PN Bowers and JR Kirby. 2010. Effects of morphological instruction on vocabulary acquisition. *Reading and Writing* 23(5):515–537.

Mirjana Bozic, William D. Marslen-Wilson, Emmanuel A. Stamatakis, Matthew H. Davis, and Lorraine K. Tyler. 2007. Differentiating morphology, form, and meaning: Neural correlates of morphological complexity. *Journal of Cognitive Neuroscience* 19(9):1464–1475.

Kate Cain, Andrea S. Towse, and Rachael S. Knight. 2009. The development of idiom comprehension: An investigation of semantic and contextual processing skills. *Journal of Experimental Child Psychology* 102(3):280–298.

J Chmielik and N Grabar. 2011. Détection de la spécialisation scientifique et technique des documents biomédicaux grâce aux informations morphologiques. *TAL* 51(2):151–179.

J Cohen. 1960. A coefficient of agreement for nominal scales. *Educational and Psychological Measurement* 20(1):37–46.

RA Côté. 1996. *Répertoire d'anatomopathologie de la SNOMED internationale, v3.4.* Université de Sherbrooke, Sherbrooke, Québec.

AP Dempster, NM Laird, and DB Rubin. 1977. Maximum likelihood from incomplete data via the em algorithm. *Journal of the Royal Statistical Society* 39(1):1–38.

Petra Dohmes, Pienie Zwitserlood, and Jens Bölte. 2004. The impact of semantic transparency of morphologically complex words on picture naming. *Brain and Language* 90(1-3):203–212.

Laurie Beth Feldman and Emily G. Soltano. 1999. Morphological priming: The role of prime duration, semantic transparency, and affix position. *Brain and Language* 68(1-2):33–39.

Laurie Beth Feldman, Emily G Soltano, Matthew J Pastizzo, and Sarah E Francis. 2004. What do graded effects of semantic transparency reveal about morphological processing? *Brain and Language* 90(1-3):17–30.

Douglas Fisher. 1987. Knowledge acquisition via incremental conceptual clustering. *Machine Learning* 2(2):139–172.

R Flesch. 1948. A new readability yardstick. *Journ Appl Psychol* 23:221–233.

T François and C Fairon. 2013. Les apports du TAL à la lisibilité du français langue étrangère. *TAL* 54(1):171–202.

S Frisson, E Niswander-Klement, and A Pollatsek. 2008. The role of semantic transparency in the processing of english compound words. *Br J Psychol* 99(1):87–107.

N Gala, T François, and C Fairon. 2013. Towards a french lexicon with difficulty measures: NLP helping to bridge the gap between traditional dictionaries and specialized lexicons. In *eLEX-2013*.

L Goeuriot, N Grabar, and B Daille. 2007. Caractérisation des discours scientifique et vulgarisé en français, japonais et russe. In *TALN*. pages 93–102.

N Grabar, S Krivine, and MC Jaulent. 2007. Classification of health webpages as expert and non expert with a reduced set of cross-language features. In *AMIA*. pages 284–288.

L Guilbert. 1971. De la formation des unités lexicales. In Paris Larousse, editor, *Grand Larousse de la langue française*, pages IX–LXXXI.

R Gunning. 1973. *The art of clear writing.* McGraw Hill, New York, NY.

Henning Holle, Thomas C Gunter, and Dirk Koester. 2010. The time course of lexical access in morphologically complex words. *Neuroreport* 21(5):319–323.

C Iacobini. 2003. Composizione con elementi neoclassici. In Maria Grossmann and Franz Rainer, editors, *La formazione delle parole in italiano*, Walter de Gruyter, pages 69–96.

Gonia Jarema, Céline Busson, Rossitza Nikolova, Kyrana Tsapkini, and Gary Libben. 1999. Processing compounds: A cross-linguistic study. *Brain and Language* 68(1-2):362–369.

Dirk Koester and Niels O. Schiller. 2011. The functional neuroanatomy of morphology in language production. *NeuroImage* 55(2):732–741.

T Kohonen. 1989. *Self-Organization and Associative Memory*. Springer.

D Kokkinakis and M Toporowska Gronostaj. 2006. Comparing lay and professional language in cardiovascular disorders corpora. In Australia Pham T., James Cook University, editor, *WSEAS Transactions on BIOLOGY and BIOMEDICINE*. pages 429–437.

LJ Kuo and RC Anderson. 2006. Morphological awareness and learning to read: A cross-language perspective. *Educational Psychologist* 41(3):161–180.

G Leroy, S Helmreich, J Cowie, T Miller, and W Zheng. 2008. Evaluating online health information: Beyond readability formulas. In *AMIA 2008*. pages 394–8.

Gary Libben, Martha Gibson, Yeo Bom Yoon, and Dominiek Sandra. 2003. Compound fracture: The role of semantic transparency and morphological headedness. *Brain and Language* 84(1):50–64.

Heidi Lüttmann, Pienie Zwitserlood, and Jens Bölte. 2011. Sharing morphemes without sharing meaning: Production and comprehension of german verbs in the context of morphological relatives. *Canadian Journal of Experimental Psychology/Revue canadienne de psychologie expérimentale* 65(3):173–191.

A McCallum, K Nigam, and LH Ungar. 2000. Efficient clustering of high dimensional data sets with application to reference matching. In *ACM SIGKDD international conference on Knowledge discovery and data mining*. pages 169–178.

Deborah McCutchen, Sara Stull, Becky Logan Herrera, Sasha Lotas, and Sarah Evans. 2014. Putting words to work: Effects of morphological instruction on children's writing. *J Learn Disabil* 47(1):1–23.

Marcus Meinzer, Aditi Lahiri, Tobias Flaisch, Ronny Hannemann, and Carsten Eulitz. 2009. Opaque for the reader but transparent for the brain: Neural signatures of morphological complexity. *Neuropsychologia* 47(8-9):1964–1971.

T Miller, G Leroy, S Chatterjee, J Fan, and B Thoms. 2007. A classifier to evaluate language specificity of medical documents. In *HICSS*. pages 134–140.

F Namer. 2000. FLEMM : un analyseur flexionnel du français à base de règles. *Traitement automatique des langues (TAL)* 41(2):523–547.

Fiammetta Namer and Pierre Zweigenbaum. 2004. Acquiring meaning for French medical terminology: contribution of morphosemantics. In *Annual Symposium of the American Medical Informatics Association (AMIA)*. San-Francisco.

M Poprat, K Markó, and U Hahn. 2006. A language classifier that automatically divides medical documents for experts and health care consumers. In *MIE 2006 - Proceedings of the XX International Congress of the European Federation for Medical Informatics*. Maastricht, pages 503–508.

H Schmid. 1994. Probabilistic part-of-speech tagging using decision trees. In *International Conference on New Methods in Language Processing*. pages 44–49.

M Schonlau, L Martin, A Haas, KP Derose, and R Rudd. 2011. Patients' literacy skills: more than just reading ability. *J Health Commun* 16(10):1046–54.

Y Wang. 2006. Automatic recognition of text difficulty from consumers health information. In IEEE, editor, *Computer-Based Medical Systems*. pages 131–136.

I.H. Witten and E. Frank. 2005. *Data mining: Practical machine learning tools and techniques*. Morgan Kaufmann, San Francisco.

Q Zeng-Treiler, H Kim, S Goryachev, A Keselman, L Slaugther, and CA Smith. 2007. Text characteristics of clinical reports and their implications for the readability of personal health records. In *MEDINFO*. Brisbane, Australia, pages 1117–1121.

Entity-Centric Information Access with the Human-in-the-Loop for the Biomedical Domains

Seid Muhie Yimam[†], Steffen Remus[†], Alexander Panchenko[†],
Andreas Holzinger[‡], and Chris Biemann[†]

[†]Language Technology Group, Department of Informatics
Universität Hamburg, Germany
[‡]Research Unit HCI-KDD
Institute for Medical Informatics, Statistics and Documentation
Medical University Graz, Austria
`{yimam,remus,panchenko,biemann}@informatik.uni-hamburg.de`
`a.holzinger@hci-kdd.org`

Abstract

In this paper, we describe the concept of entity-centric information access for the biomedical domain. With entity recognition technologies approaching acceptable levels of accuracy, we put forward a paradigm of document browsing and searching where the entities of the domain and their relations are explicitly modeled to provide users the possibility of collecting exhaustive information on relations of interest. We describe three working prototypes along these lines: NEW/S/LEAK, which was developed for investigative journalists who need a quick overview of large leaked document collections; STORYFINDER, which is a personalized organizer for information found in web pages that allows adding entities as well as relations, and is capable of personalized information management; and adaptive annotation capabilities of WEBANNO, which is a general-purpose linguistic annotation tool. We will discuss future steps towards the adaptation of these tools to biomedical data, which is subject to a recently started project on biomedical knowledge acquisition. A key difference to other approaches is the centering around the user in a Human-in-the-Loop machine learning approach, where users define and extend categories and enable the system to improve via feedback and interaction.

1 Introduction

Recently, knowledge management as a field faced several challenges. On one hand, sophisticated technologies and standards were developed to support knowledge-based modeling, such as domain ontologies including Disease Ontology, MeSH, and Gene Ontology[1] and the Semantic Web description languages and infrastructures including RDF, OWL, SPARQL and others[2]. On the other hand, the current approaches face three major issues: (1) knowledge bottleneck: resources required for knowledge management such as domain ontologies are not available for many domains and languages; (2) the overall approach of knowledge management did not get widely spread due to the fact that it imposes a large burden on the user, such as annotation or expertise with complex tools such as Protégé[3]; (3) modeling entire domains as large as the medical domain with (English-oriented) knowledge resources does not meet requirements of users, who are mostly specializing in a certain sub-field and also need to operate in their local language.

We propose to reload this traditional heavyweight *top-down* knowledge management approach and replace it with a much simpler and practical problem-oriented *bottom-up* approach. We choose the biomedical domain as the area of interest for our planning. Medical researchers have to process enormous amounts of literature – PubMed[4] adds about half a million papers to its index each year. Literature search and reason-

---

[1] `http://do-wiki.nubic.northwestern.edu;`
`http://ncbi.nlm.nih.gov/mesh; http://geneontology.org`
[2] `http://www.w3.org/standards/semanticweb`
[3] `http://protege.stanford.edu`
[4] `http://www.ncbi.nlm.nih.gov/pubmed`

Proceedings of the Biomedical NLP Workshop associated with RANLP 2017, pages 42–48,
Varna, Bulgaria, 8 September 2017.

ing is demanding, because of the need to reveal and maintain many complex relationships between numerous sets of entities. In order to alleviate the efforts of biomedical research related to literature we propose a novel conception to information management based on *bottom-up* construction of a problem-oriented ontology, called entity graph (EG) in this paper. Entity graphs provide a new tool for medical researchers that (1) help to document relations between biomedical entities in a compact intuitive and interpretable form; (2) generate new relations in a semi-automatic way based on corpus analysis; (3) communicate new biomedical knowledge in a form of an easily interpretable interactive graph and (4) share knowledge and annotations amongst researchers.

2 Related Work

An early conception of a system for personal information management was Memex (Bush, 1945). The proposed design suggested that all documents of a person should be indexed to be easily accessible for consultation and for sharing with other people. Several decades later, the Web and social networks implement this vision yet only partially. According to Davenport (1994), Knowledge Management (KM) is a process of capturing, distributing, and effectively using knowledge. According to Gruber (1995), an ontology is an explicit specification of conceptualization. Studer et al. (1998) defines ontology as a formal, explicit specification of shared conceptualization. Multiple other informal and formal definitions of ontology are presented by Cimiano et al. (2014). Here "conceptualization" is a worldview, a system of conceptions and their relations.

Ontologies can be either general or domain-specific. Today's content management systems are largely accessed with faceted search, i.e. with taxonomically organized vocabularies forming a semantic facet. Users of the system must learn the vocabulary in order to assign the correct terms to newly ingested documents and to perform effective searches. The Cyc project (Lenat and Guha, 1990) was an early ontology-driven attempt to model world knowledge. Jurisica et al. (1999) presented an overview of using ontologies for information management. Later, knowledge management using ontologies was driven by the Semantic Web vision (Berners-Lee et al., 2001). This eventually led to the Linked Open Data cloud of resources, containing a comprehensive collection of interlinked ontologies. One limiting factor of widespread usage of ontologies is the heavy burden of their manual construction: all concepts, attributes and relations in ontologies are added and updated manually. Moreover, even if suitable ontologies for a target domain exist, they do not come with mechanisms to recognize their concepts in unstructured text, motivating approaches that learn ontologies from text (see Biemann (2005) and Buitelaar et al. (2005)).

Both EGs and ontologies aim at providing a shared explicit conceptualization of a certain domain. However, there are several important differences between these two resources. First, EGs are task- and/or problem-specific descriptions of a domain, while ontologies are usually designed as generic knowledge representations for a given domain. Ontologies are commonly developed as general-purpose resources that are supposed to model a certain domain without taking into account specific needs of certain application. This leads in practice to the fact that most resources should be specifically tailored to fit the need of the given task, problem or application. Along these lines, Hirst (2014) notes that the worldview captured in ontologies is based on the author of the ontology, not on the user, and the knowledge is not contextualized. We argue that this is one of the key reasons of only moderate success of ontology-based knowledge management after 15 years of development. Our approach will tackle this shortcoming: entity graphs are a knowledge representation tool that is designed to be strictly task-oriented. Such a graph would contain only concepts and relations relevant to the described problem at hand omitting any irrelevant details.

Mind maps (MMs) are visual diagrams that help to organize information about certain topics. Entity graphs have several common aspects with Mind Maps and similar knowledge management structures, such as concept maps and conceptual diagrams (Willis and Miertschin, 2006; Eppler, 2006), but are not confined to a tree structure, hence they are more apt for sharing and bring provenance in documents into the representation.

BEST is a biomedical entity search tool for knowledge discovery from biomedical literature (Lee et al., 2016). Although PubMed (the free public interface to MEDLINE, which provides access to bibliographic information in MEDLINE as well as additional life science journals) provides a starting point to researchers, it only provides lists of relevant articles, leaving the task of extracting required information to the researchers themselves. Existing context extraction systems have limitations, such as 1) they provide outdated or incomplete results 2) the processing takes longer, and 3) most of them depend on conventional search system structures to return relevant information. BEST is developed to face the challenges of getting relevant documents from biomedical literature publications, addressing most challenges by directly returning ten relevant entities for a user's query instead of a list of documents. Our approach differs from BEST in many aspects such as 1) instead of relying on existing entity dictionaries, we use a semi-supervised entity recognition system, 2) instead of returning a pre-computed list of (indexed) results, our approach directs the researcher in pinpointing the required information with directed visual exploration, i.e. a guided search, 3) in addition to pre-defined entity types or dictionaries, our approach allows researchers to define their own entity types without the need of advanced pre-processing or text mining knowledge, i.e. adaptive annotation.

Zhang and Elhadad (2013) propose an unsupervised approach for detecting biomedical entities. Instead of hand-crafted rules or annotated dataset, this work first identifies classes of entities based on UMLS[5] semantic groups in order to collect seed terms. Next, they extract chunks in order to automatically determine named entity boundaries. Finally, they use a similarity based approach to automatically group named entities into specific semantic classes. While this approach is beneficial to identify biomedical entities, it has some drawbacks compared to our approach: 1) their approach depends on the collection of seed terms, 2) it assumes that every biomedical document is available at all times.

3 Three Technologies for Entity-Centric Information Access

While we target the biomedical domain, we will describe our previous work on other domains. The entity types might change, but the principles of the entity graph is transferable across domains.

3.1 Adaptively Annotating Entities with WEBANNO

Supervised named entity recognition (NER) systems require a substantial amount of annotated data to achieve high quality performance. We present an interactive and adaptive annotation approach. Instead of using a large sets of general purpose annotation corpora, we focus on specifically collecting high quality sets of in-domain annotations. In a case study for adaptive biomedical entity annotation, we used the automation component of WEBANNO, which is a web-based annotation tool with an online machine learning component (Yimam et al., 2014). Annotations are created in an interactive and incremental approach. The process is interactive in such a way that the tool suggests annotations that can be accepted, rejected or corrected by the annotator, whereby machine learning model gets better in time.

3.2 Case Study: Entity Annotation

We conducted an annotation task for identifying medical entities using WEBANNO automation, which is focused on B-Chronic lymphocytic leukemia (B-CLL). A medical expert selects domain related abstracts for annotation. Unlike previous approaches, the expert starts annotating texts without prior determination of the entity types. During the annotation process, important entities are identified that could help retrieving relevant documents about B-CLL. In a first step, we annotated five abstracts and use them for training to produce suggestions.

The following entity types are identified throughout the task: CELL, CONDITION, DISORDER, GENE, MOLECULE, PROTEIN, MOLECULAR PATHWAY and SUBSTANCE. We can see the following advantages of the adaptive annotation approach: 1) it makes the annotation task faster by producing correct predictions after annotating only a few number of documents, 2) the process helps the annotator to determine entity types unlike traditional approaches where the types are predefined by experts beforehand. This makes the identification of entity types more complete and robust (see details in Yimam et al., 2016a).

One of the typical relations between biomedical entities describe the cause and effect of diseases. Again, supervised machine learning approaches for automatic relation extraction requires more ef-

---

[5]Unified Medical Language System (UMLS) is a widely used ontology of biomedical terms available at `https://www.nlm.nih.gov/research/umls/`.

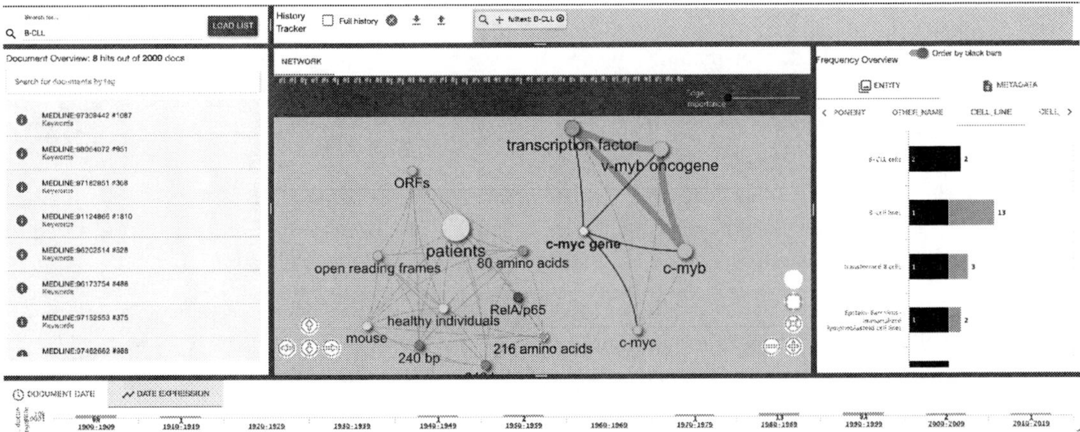

Figure 1: NEW/S/LEAK UI overview of the GENIA term annotation corpus. The example shows a B-CLL query and the graph shows involved DNA regions, "c-myc gene" is selected.

fort. For rapid annotation of relations, the relation copy annotator in WEBANNO was used, where relation suggestions are provided as soon as annotators create the first relation annotations. This functionality has the following advantages: a) experts can annotate entities as well as relation annotations at the same time, b) instances of the same entity and relation are automatically suggested for the running document as well as other unfinished documents.

3.3 Collection Insights with NEW/S/LEAK

NEW/S/LEAK is a tool designed to support investigative data journalism by exploring large sets of input documents, typically leaked documents (Yimam et al., 2016b). Named entities, such as persons, organizations, and locations, are automatically identified and ranked by importance. A global graph of entities is constructed, which is subsequently used to display high-level interactions among those entities. The tool is intended to guide investigative data journalists, by offering a rich set of possible interactions, among which are: full text search, entity merger or removal, document aggregation using meta-data, and many more.

Journalists, as targeted user group, can browse the document collection using the interactive interface (see Figure 1). It enables faceted document exploration within several views: 1) the **graph view** shows named entities and their relations, 2) the **document timeline view** shows document frequency in different epochs, 3) the **document view** is composed of the document list and a document text for reading, and 4) the **metadata views** include the search- and history views, which offer different metadata for filtering relevant or irrelevant documents.

The views are interactive, i.e. users can browse and explore the document collection on demand. The user starts with exploring entities and their connections in the graph view or by searching for entities and keywords. All interactions in the views define a filter that constrains the current document set, which in turn changes the displayed information content. User-selected entities are highlighted in the documents.

Graph view: entities and their co-occurrences The graph view shows a set of entities as nodes and their connections as links. The node size denotes the frequency of an entity, the node color denotes the entity type. The number of shown entities can be set by the user individually for each facet (entity type). The edge thickness and label denotes the size and relation of co-occurrence of the involved entities within the documents.

Document timeline The document timeline lists the number of documents in a specific epoch. Users can refine their search to see the document distribution over years, months or days.

Document view The document view shows a list of documents with their heading as selected by the currently active filters. For large document collections, the documents are loaded on demand. The document text view shows the text of the document, where the entities displayed in the graph are highlighted and underlined. The underline color corresponds to the type of entity. Selected entities in the graph are highlighted, which enables a "close reading" mode to verify hypotheses formed in the so-called "distant reading" visualization (Moretti, 2007).

Metadata, search and history tracing This view is mainly used to filter documents based on different criteria such as metadata, entities, search terms/key words, etc. The history tracer helps the journalist to modify the search facets.

3.4 Personalized Knowledge Management with STORYFINDER

STORYFINDER is a toolkit that aims to keep information managed which is found and processed while browsing the web (Remus et al., 2017). The major goal is to organize a personal history of *bits of information* in form of entities and their relations rather than a history of web pages while still being able to find the source of a particular information bit in the respective web pages.

The system consists of three major components (cf. Fig. 2): 1) the Mozilla Firefox **browser plugin**, which: listens and reacts to a user's actions; initiates the analysis of a currently visited webpage on the backend server; and provides a side pane view to visualize the collected information; 2) the **server backend**, which: performs the analysis of a webpage; extracts metadata and stores the information for later access; and 3) the **interactive web page**, which: provides real-time access to the new information and is embedded in the plugin's side pane and can be accessed as a regular web page too.

In its current form, STORYFINDER is targeted for processing news texts; it automatically extracts *named entities* and draws an edge in a knowledge graph representation if two distinct entities co-occur in the same sentence (Fig. 3a).

The entities are subsequently highlighted within the current article for better visual appearance (Fig. 3b). The graph, i.e. the entities as nodes and their relations as edges are fully editable (Fig. 3c).

Due to the modular REST architecture regarding the NLP components within the backend server, every automatic component is exchangeable, e.g. in order to automatically identify medical entities such as proteins, we merely need a reliable protein tagger. In order to build such a tagger, annotated data is needed, which calls for an integration with adaptive annotation(Section 3.1).

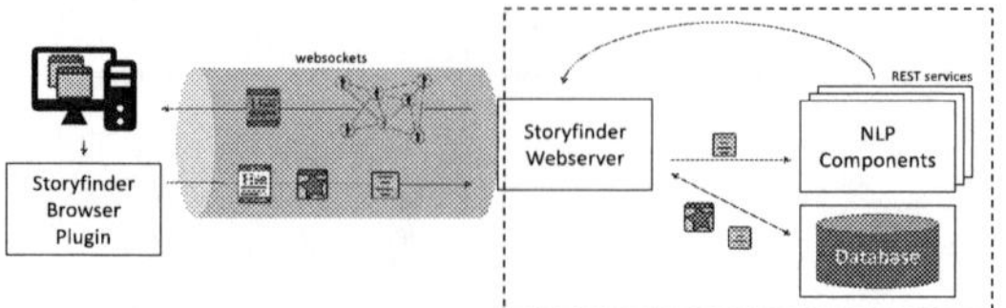

Figure 2: Schema of STORYFINDER's components: The browser plugin, the server backend, and the interactive web page.

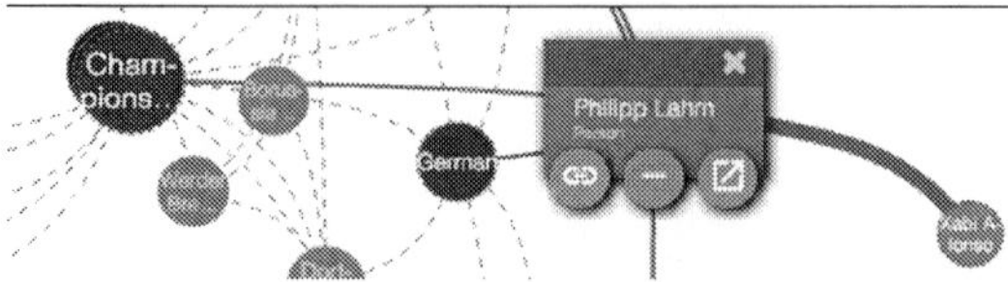

(a) The entity 'Philipp Lahm' is selected, other nodes and edges are grayed out except direct neighboring edges and nodes. Additionally an edge is hovered (rightmost thick edge).

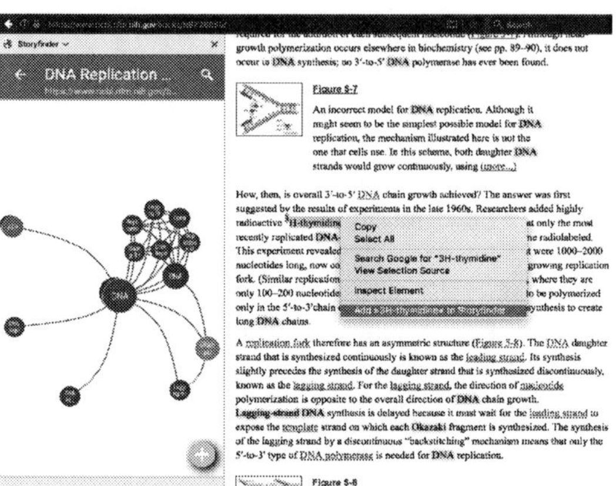

(b) Screenshot of the default STORYFINDER plugin view. A currently visited webpage is analyzed, and the extracted entities are highlighted in an overlay. Entities are rendered in a graph together with their relations in the STORYFINDER webpage, which is shown in a side pane of the browser.

(c) Manually adding entities of arbitrary kind can be accomplished via the plugin by right clicking any term or phrase.

Figure 3: Selected STORYFINDER screenshots.

4 Towards Information Management with Human-in-the-Loop

Within our newly started project, we will implement a prototype that uses the entity graph representation as the primary means for visualizing and

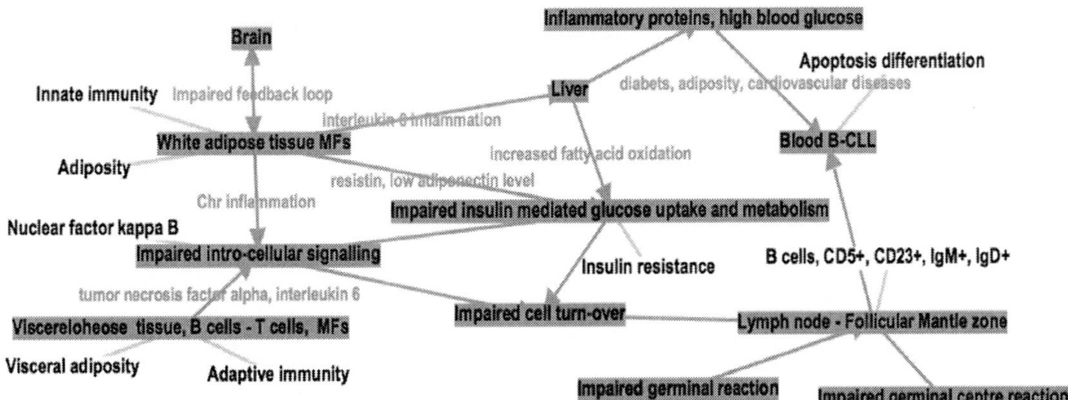

Figure 4: An entity graph summarizing the literature research on B-CLL. The key symptoms, drugs and treatments around the B-CLL are shown with their labeled relations. From labels, it becomes clear how entities relate to the topic, click on edges retrieves documents where connected concepts co-occur.

accessing biomedical research documents, integrating elements from prototypes described above. Key to the approach is to think the user in the center of the process and offer the user an adaptive ML environment (Holzinger, 2016; Holzinger et al., 2017) where manual effort in terms of annotating entities or classifying relations immediately pays of in an improved representation in the EG. To exemplify how this could look like, Figure 4 shows an example from leukemia research. Entities and their relations have been annotated and semi-automatically recognized in a personal collection of MEDLINE papers (Yimam et al., 2016a). Interacting with the network allows to find respective documents.

With these actions, the biomedical researcher can utilize the entity graph as a visually supportive notepad. Note that this goes well beyond a traditional notepad since collections of properties of entities usually do not get linked, and this also goes well beyond creativity tools such as e.g. mind maps, since it does not only displays concepts, but facilitates linking to source documents. Note further that while automatic methods aid the process, the biomedical researcher is in full control of the entity graph and can correct errors in the automatic processing in case they are relevant for the question of investigation.

Last, but not least, the individual entity graphs can be merged into a global structure by sharing among researchers. Thus in our approach, the conceptualization of a domain will be modeled from BOTTOM-UP, and not from TOP-DOWN as in the traditional knowledge management approach. Therefore, collaborative efforts of the crowd will lead to construction of a global entity graph of a domain in an incremental and problem-driven way. The global graph can be used to softly suggest edge annotations while a user constructs a new graph, making the overall process of entity graph construction backed up by a huge global entity graph, which has provenance information (i.e. who has entered information, based on which document) for mutual understanding. The global graph can also incorporate information from resources, such as MeSH, Gene and Disease Ontology. Challenges in the adaptation include a high-quality tagging of biomedical entities, preprocessing such as dependency parsing for relevant languages, the design of the user interface and a responsive online-adaptive machine learning model.

5 Conclusion

We proposed a new schema for entity-centric information extraction and -access for biomedical entities. We highlighted current drawbacks and new challenges, and presented existing tools for information extraction (WEBANNO), visualization and navigation (NEW/S/LEAK), and personalized information and knowledge management (STORYFINDER), which all together can be combined, adapted, and re-focused in order to provide a data driven, *bottom-up*, conceptualization approach. Here, the *Human-in-the-Loop* is an integral component, where not only the machine learning models for information extraction are supported and improved by users over time, the final entity graph becomes larger, cleaner, more precise and thus more usable for the users.

Acknowledgments

This research was supported by the Federal Ministry for Education and Research (Germany) under grant no. 01DS17033 and by the Volkswagen Foundation under grant no. 90 847.

References

Tim Berners-Lee, James Hendler, and Ora Lassila. 2001. The Semantic Web. *Scientific American* 284(5):28–37.

Chris Biemann. 2005. Ontology learning from text - a survey of methods. *LDV Forum* (2):75–93.

Paul Buitelaar, Philipp Cimiano, and Bernardo Magnini. 2005. *Ontology learning from text : methods, evaluation and applications*, volume 123. IOS Press.

Vannevar Bush. 1945. As we may think. *The Atlantic Monthly* 176(1):101–108.

Philipp Cimiano, Christina Unger, and John McCrae. 2014. Ontology-based interpretation of natural language. *Synthesis Lectures on Human Language Technologies*. 7.2:1–178.

Thomas H. Davenport. 1994. Saving it's soul: Human-centered information management. *Harvard Business Review* 72(2):119–131.

Martin J. Eppler. 2006. A comparison between concept maps, mind maps, conceptual diagrams, and visual metaphors as complementary tools for knowledge construction and sharing. *Information Visualization* 5:202–210.

Thomas R. Gruber. 1995. Toward principles for the design of ontologies used for knowledge sharing. *Int. J. Hum.-Comput. Stud.* 43(5-6):907–928.

Graeme Hirst. 2014. *Overcoming Linguistic Barriers to the Multilingual Semantic Web*, Berlin, Heidelberg, pages 3–14.

Andreas Holzinger. 2016. Interactive machine learning for health informatics: when do we need the human-in-the-loop? *Brain Informatics* 3(2):119–131.

Andreas Holzinger, Markus Plass, Katharina Holzinger, Gloria Cerasela Crisan, Camelia-M. Pintea, and Vasile Palade. 2017. A glass-box interactive machine learning approach for solving NP-hard problems with the human-in-the-loop. *ArXiv e-prints* .

Igor Jurisica, John Mylopoulos, and Eric Yu. 1999. Using ontologies for knowledge management: An information systems perspective. In *Proceedings of the ASIS Annual Meeting*. Washington, DC, USA, pages 482–496.

Sunwon Lee, Donghyeon Kim, Kyubum Lee, Jaehoon Choi, Seongsoon Kim, Minji Jeon, Sangrak Lim, Donghee Choi, Sunkyu Kim, Aik-Choon Tan, and Jaewoo Kang. 2016. Best: Next-generation biomedical entity search tool for knowledge discovery from biomedical literature. *PLOS ONE* 11(10):1–16.

Douglas Lenat and Ramanathan V. Guha. 1990. *Building Large Knowledge Bases*. Addison-Wesley Pub. Co, Reading, MA.

Franco Moretti. 2007. *Graphs, maps, trees : abstract models for a literary history*. Verso, London, UK.

Steffen Remus, Manuel Kaufmann, Kathrin Ballweg, Tatiana von Landesberger, and Chris Biemann. 2017. Storyfinder: Personalized knowledge base construction and management by browsing the web. In *Proceedings of the 26th ACM International Conference on Information and Knowledge Management*. Singapore, Singapore. To appear.

Rudi Studer, V. Richard Benjamins, and Dieter Fensel. 1998. Knowledge engineering: Principles and methods. *Data Knowl. Eng.* 25(1-2):161–197.

Cheryl L. Willis and Susan L. Miertschin. 2006. Mind maps as active learning tools. *Journal of computing sciences in colleges* 21(4):266–272.

Seid Muhie Yimam, Chris Biemann, Ljiljana Majnaric, Sefket Sabanovic, and Andreas Holzinger. 2016a. An adaptive annotation approach for biomedical entity and relation recognition. *Brain Informatics* 3(3):157–168.

Seid Muhie Yimam, Richard Eckart de Castilho, Iryna Gurevych, and Chris Biemann. 2014. Automatic annotation suggestions and custom annotation layers in WebAnno. In *Proc. of ACL 2014: System Demonstrations*. Baltimore, MD, USA, pages 91–96.

Seid Muhie Yimam, Heiner Ulrich, Tatiana von Landesberger, Marcel Rosenbach, Michaela Regneri, Alexander Panchenko, Franziska Lehmann, Uli Fahrer, Chris Biemann, and Kathrin Ballweg. 2016b. new/s/leak – information extraction and visualization for investigative data journalists. In *Proceedings of ACL-2016 System Demonstrations*. Association for Computational Linguistics, Berlin, Germany, pages 163–168.

Shaodian Zhang and Noémie Elhadad. 2013. Unsupervised biomedical named entity recognition. *J. of Biomedical Informatics* 46(6):1088–1098.

One model per entity: using hundreds of machine learning models to recognize and normalize biomedical names in text

Victor Bellon
MINES ParisTech,
PSL-Research University,
CBIO-Centre de bio-informatique
`victor.bellon@mines-paristech.fr`

Raul Rodriguez-Esteban
Roche Innovation Center Basel
`raul.rodriguez-esteban@roche.com`

Abstract

We explored a new approach to named entity recognition based on hundreds of machine learning models, each trained to distinguish a single entity, and showed its application to gene name identification (GNI). The rationale for our approach, which we named "one model per entity" (OMPE), was that increasing the number of models would make the learning task easier for each individual model. Our training strategy leveraged freely-available database annotations instead of manually-annotated corpora. While its performance in our proof-of-concept was disappointing, we believe that there is enough room for improvement that such approaches could reach competitive performance while eliminating the cost of creating costly training corpora.

1 Background

Recognizing names in text is a longstanding task in natural language processing (NLP) known as named-entity recognition (NER). In biomedical text mining (or BioNLP), the focus of NER is on certain technical names (terms) such as those of chemical compounds, genes, species and anatomical parts. Recognizing such names alone, however, is of limited application as, in practice, they often need to be linked to other facts. This can be done by first mapping them to unique name identifiers—a task known as normalization or grounding. Recognizing and normalizing names of genes and gene products, in particular, has drawn much attention from the BioNLP community (Leser and Hakenberg, 2005). These names are usually considered a single class of terms due to their overlapping vocabularies (Hatzivassiloglou et al., 2001). Thus, here we refer to them as simply gene names.

The tasks of gene name recognition (GNR) and gene name normalization (GNN) involve, respectively, the recognition and normalization of gene names found in text. Gene name identification (GNI) is the combination of gene name recognition and normalization (GNR + GNN) (see the framework by Krauthammer and Nenadic (2004)). State-of-the-art GNI methods involve machine learning algorithms, such as conditional random fields (CRF), trained under supervised learning. Supervised learning requires gold-standard training and testing sets, which for GNI typically are sets of documents (corpora) that have been manually annotated for gene names by expert curators. Several community challenges have been organized to foster the improvement of GNI algorithms (Morgan et al., 2008; Lu et al., 2011). However, despite such efforts, even the best algorithms suffer from an important weakness. Namely that their performance has been shown to degrade outside of their training and testing corpora, decaying to levels barely above those of rule-based systems involving dictionary-matching rules together with filtering of noisy names (Rebholz-Schuhmann et al., 2013; Rodriguez-Esteban, 2016b,a).

It has been suggested that the shortcomings of current GNI machine learning algorithms could be addressed in two ways: (1) by training models with different, diverse corpora (Rebholz-Schuhmann et al., 2013) (see, in that respect, the work of Kaewphan et al. (2016) with cell line names), and (2) by using "domain adaptation" techniques, which consist in adapting machine learning models to the characteristics of the input text. The limited size and number of existing gold-standard corpora, and the cost of creating new ones, represent, however, a bottleneck for (1). For (2), experiments with domain adaptation in biomedical text have led, thus far, only to modest improvements in performance (Miwa et al., 2012).

Here we describe an alternative approach that

Proceedings of the Biomedical NLP Workshop associated with RANLP 2017, pages 49–54,
Varna, Bulgaria, 8 September 2017.

can be applied to problems that require the identification of large but finite sets of entities, particularly in biomedicine. To begin with, instead of using a gold-standard corpus as training set, we propose utilizing the wealth of manual annotations that currently exist in biomedical databases. Indeed, several freely-available databases provide a growing number of annotations concerning gene name identifiers associated to biomedical documents. The main drawback of these annotations is that they are weakly labeled, as they do not specify the precise location in which the genes are mentioned within the documents. However, there are ways to infer these locations (Jain et al., 2016).

While past GNI studies have not leveraged annotations from biomedical databases, there are examples of their use for GNN (Wermter et al., 2009; Zwick, 2015; Chen et al., 2015). In these studies contextual features were created out of the annotated biomedical documents to resolve ambiguous gene mentions. Besides for GNN, weakly-labeled database annotations have been used in BioNLP for identifying protein-specific residues (Ravikumar et al., 2012) and annotating Medline abstracts with Gene Ontology terms (Gobeill et al., 2013). In another example, Furrer et al. (2014) used a biomedical database called BioGRID (Chatr-Aryamontri et al., 2015) for the purpose of training and testing an algorithm for extracting protein-protein interactions (PPI).

Leveraging database annotations for GNI is not straightforward. We have implemented our approach in a way that, as far as we can tell, has not been described in the NER literature before (biomedical or otherwise). Our method involves training many machine learning models, each model trained to identify a single entity (i.e. a single gene) rather than, as it is commonplace, training one or a handful of models to identify all entities. We call this approach "one model per entity" (OMPE).

2   Methods

As building block for our OMPE system we used BANNER (Leaman and Gonzalez, 2008), which is a machine learning algorithm for NER built on CRF. While BANNER is based on a generic model that can be trained to identify any class of terms, it has shown state-of-the-art performance in GNR (Kabiljo et al., 2009). Our strategy consists in using multiple BANNER models, each model being

responsible for detecting the mentions of a single gene. That means that a gene name mention that is recognized by a BANNER model can be automatically mapped to the gene for which the model was trained.

2.1   Training set

To create our training set we built first a database of positive training examples containing sentences that mention gene names. Each sentence in the database was associated to a gene identifier (NCBI Gene ID), corresponding to a gene mentioned in the sentence, and to a document identifier (PubMed ID), corresponding to the document source of the sentence. The {NCBI Gene ID, PubMed ID} pairs came from the following publicly-available databases: gene2pubmed, UniProt, BioGRID (Chatr-Aryamontri et al., 2015) and Gene Reference into Function (GeneRIF) (see Table 1).

Source	Genes	Documents	Mentions
gene2pubmed	34 004	493 620	1 087 465
UniProt	21 383	22 539	68 966
BioGRID	11 832	23 925	66 358
GeneRIF	17 462	386 927	641 354

Table 1: Statistics of the different datasets used.

Because these databases do not specify the location of the gene mentions in the source documents, we retrieved each source document from the Medline baseline 2015 and attempted to find their locations. In order to do that we leveraged gene names and synonyms from the NCBI Gene database. This database, however, does not include all the gene name variations and synonyms that authors use in practice (Hirschman et al., 2002; Liu et al., 2006). To increase recall we therefore expanded the list of gene names and synonyms following Schuemie et al. (2007). By using this expanded list to look up gene names in the source documents we created a set of positive examples for each of the genes annotated in the aforementioned databases.

For training (and testing) we only considered genes for which we had at least 32 positive examples (this cut-off was a compromise between coverage and amount of training data available), which totaled 2180 with a median of 281 positive examples per gene. These genes covered approximately 80% of all gene mentions appearing in the test corpus.

Negative training examples were selected according to different strategies. First, we created certain modified versions of the positive examples. Modifications consisted in the deletion of words within the gene names that made reference to a certain function, such as *receptor, inhibitor, enhancer*. For example, while *TNF-α receptor* refers to gene ID *7132*, *TNF-α* corresponds to gene ID *7124*.

The second type of negative examples that we selected consisted in positive examples belonging to genes that share synonyms. For example, the gene name *FAT* may refer to gene ID *2195* or *948*. Thus, positive examples for gene *948* can be used as negative examples for gene *2195*. Finally, we included as negative examples randomly selected sentences from the English Wikipedia (not from any particular domain) and positive examples from randomly selected genes.

2.2 Test set

For testing the performance of our OMPE system we used a modified dataset based on the gold standard from the BioCreative 2 Gene Normalization (BC2GN) challenge (Morgan et al., 2008). The BC2GN training set covers 281 abstracts and 684 gene annotations, and the testing set covers 262 abstracts and 785 gene annotations. As we used an independent training set based on freely-available database annotations we could employ both BC2GN training and testing datasets to create our BC2GN$_{mod}$ test dataset.

When building the BC2GN$_{mod}$ dataset we only considered gene annotations for 621 unique genes from the 1156 genes present in the original BC2GN datasets—those for which we had more than 32 positive examples in our training database. Thus, BC2GN$_{mod}$ contained a total of 841 human gene annotations.

We compared our OMPE system against GNAT (Hakenberg et al., 2008, 2011), which is a state-of-the-art system for GNI (Rebholz-Schuhmann et al., 2013). We evaluated the prediction quality of our system according to the number of true positives (TP), false positives (FP) and false negatives (FN), and according to precision (P), recall (R) and F-measure (F).

2.3 Computation

We made use of two different computational configurations for training and testing. First, we used a server with 40 CPU cores at 2.4 GHz and 567 GB RAM. This server was used for both generating the training set and making the final predictions. As training the models is the most computationally demanding task, we used a cluster computer with 164 nodes, each node possessing 2 CPUs with 12 cores (Intel Xeon Processor E5-2680 v3) and 256 GB of memory. In this configuration the median model training time was 212 seconds.

3 Results

Two versions of the OMPE system were tested and compared against the output of GNAT. The first version (OMPE1) used the standard BANNER implementation, in which the most probable class is associated to every token. The second version (OMPE2) used a modified BANNER that required the probability of a token being a mention to be larger than a certain threshold, which we set to 0.95.

Results for predictions over the BC2GN$_{mod}$ dataset can be seen in Table 2. GNAT showed a high performance, with a recall of 0.762 and a precision of 0.881, corresponding to 892 TPs and only 121 FPs. The OMPE1 system achieved, on the other hand, a recall of 0.331 and a precision of 0.215 caused by the large number of FPs, 1413.

Method	TP	FP	FN	P	R	F
GNAT	892	121	278	.881	.762	.817
OMPE1	387	1413	783	.215	.331	.261
OMPE2	355	575	815	.382	.303	.338

Table 2: Performance of the 3 different methods.

To reduce the number of FPs we set a threshold to the probability of accepting a prediction (OMPE2). By using the threshold we dramatically reduced the number of FPs from 1413 to 575, increasing the precision to 0.382 while slightly decreasing the recall to 0.303.

In Figure 1 we show a comparison of each method's individual performance. In this figure, the first row compares OMPE1 and GNAT, while the second row compares OMPE2 and GNAT. The last row compares OMPE1 and OMPE2. Light colors represent the individual performance of the methods and dark colors the difference between them. The first and second column show the precision and recall, respectively. Genes were ordered according to performance differences between methods.

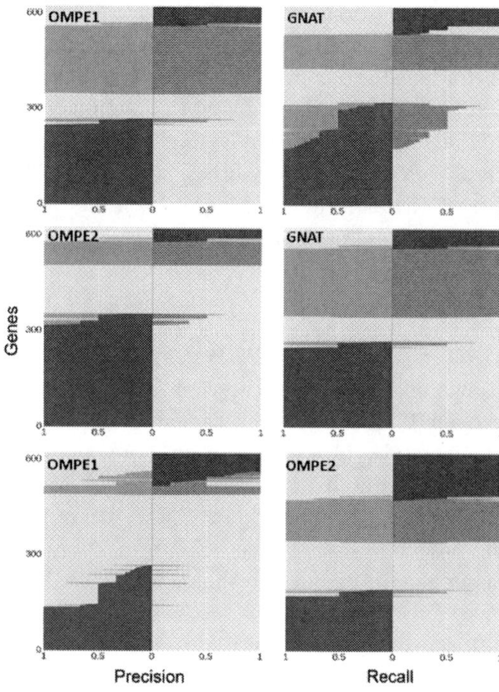

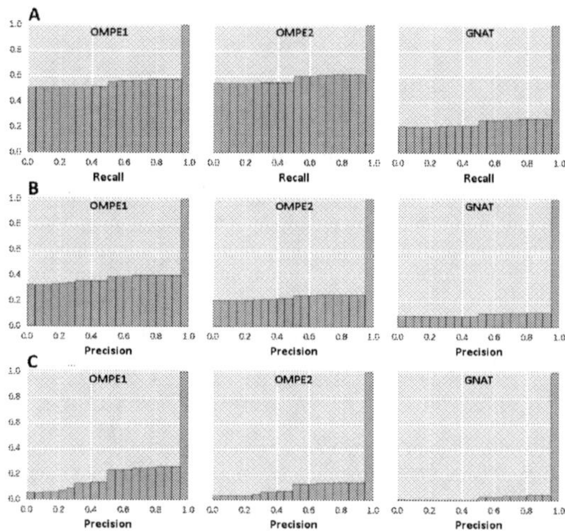

Figure 1: Difference in precision and recall of the different methods on an individual gene basis. The first row compares OMPE1 and GNAT, while the second row compares OMPE2 and GNAT. The last row compares OMPE1 and OMPE2. The first and second column show the precision and recall, respectively. Light colors represent the individual performance of the methods and dark colors the difference between them. Genes were ordered according to performance differences between methods.

Figure 1 shows that there is a set of genes in which one of the algorithms works well but the other algorithms do not. Moreover, the use of a threshold in OMPE2 leads to an increase in the precision over a large number of genes and to a decrease in only a small number of them. Figure 2, on the other hand, shows the cumulative frequency distribution of precision and recall for genes predicted in the test datasets with the different methods.

4 Discussion

An advantage of the OMPE approach is that it allows targeted performance improvements with respect to specific gene names. Positive examples and synonyms belonging to particularly challenging gene names can be modified interactively

Figure 2: Cumulative frequency distribution of genes at each precision and recall level. (A) Recall for prediction of genes in the test dataset. (B) Precision for prediction of genes in the test dataset. (C) Precision for algorithms focused on the genes known to be present in the test dataset.

without the need for retraining the entire system (all models in our case), unlike in interactive single-model approaches such as *tagtog* (Cejuela et al., 2014). Another advantage of OMPE is its robustness, as it is not trained on a particular hand-selected corpus. Thus, our results with the $BC2GN_{mod}$ corpus are not biased by the training set utilized.

A challenge for training the OMPE system is the selection of negative examples. It is important to select negative examples that are as similar as possible to the positive examples, meaning examples that are closest to the class separation boundary—analogous to what support vectors represent for support vector machines (SVMs). One of our approach's limitations is its reduced recall due to the low number of positive examples that exist for many genes. Such genes are, on the other hand, less likely to be mentioned in the biomedical literature and, as biomedical databases continue to grow, the number of positive examples for those genes will keep increasing as well.

Finally, our focus was only on human genes. The identification of genes from additional species would have required greater computational resources. An OMPE system that covered all protein-expressing genes would need to be trained for around 20 000 genes (Ezkurdia et al., 2014).

In this sense, and in the reliance on large, growing biomedical databases, our approach has a futuristic stance, meaning that it will become more feasible with time. As "Big Computing" infrastructure, such as cloud computing, becomes increasingly available and more powerful, it will become more practical to implement systems such as OMPE. It is important to stress that computational requirements differ greatly between training an OMPE system and deploying it for prediction, which requires far lower computational power.

Beyond the GNI example shown here, OMPE can be used to identify other types of entities with a finite cardinality, such as (in the BioNLP field) diseases, cell types, cell lines and anatomical parts. We have focused here on GNI because it has been already widely investigated and has multiple applications, such as the tracking of biomedical facts and trends (Cokol and Rodriguez-Esteban, 2008; Cokol et al., 2007; Rodriguez-Esteban and Loging, 2013). Beyond NER, the OMPE approach could also be applied to other classification problems in which the class cardinality is below a computationally-feasible threshold. The rationale would again be that increasing the number of models could ease ("relieve") the learning task to each individual model.

5   Conclusion

In this study we have shown a new approach for GNI that takes advantage of the decreasing costs of computing and the increasing availability of annotated data to train hundreds of machine learning models. Our proof of concept did not reach acceptable performance levels but, due to the fact that there remains ample room for potential improvements, such strategies could become competitive for GNI and other domains in the future. Following the remarks from Halevy et al. (2009) in "The unreasonable effectiveness of data," we should learn to "use available large-scale data rather than hoping for annotated data that isn't available."

References

JM Cejuela, P McQuilton, L Ponting, SJ Marygold, R Stefancsik, GH Millburn, B Rost, and FlyBase Consortium. 2014. tagtog: interactive and text-mining-assisted annotation of gene mentions in PLOS full-text articles. *Database (Oxford)* 0(bau033). https://doi.org/10.1093/database/bau033.

A Chatr-Aryamontri, B J Breitkreutz, R Oughtred, L Boucher, S Heinicke, D Chen, C Stark, A Breitkreutz, N Kolas, L O'Donnell, T Reguly, J Nixon, L Ramage, A Winter, A Sellam, C Chang, J Hirschman, C Theesfeld, J Rust, M S Livstone, K Dolinski, and M Tyers. 2015. The BioGRID interaction database: 2015 update. *Nucleic Acids Res* 43(Database issue):D470–478. https://doi.org/10.1093/nar/gku1204.

G Chen, J Zhao, T Cohen, C Tao, J Sun, H Xu, E V Bernstam, A Lawson, J Zeng, A M Johnson, V Holla, A M Bailey, H Lara-Guerra, B Litzenburger, F Meric-Bernstam, and W Jim Zheng. 2015. Using ontology fingerprints to disambiguate gene name entities in the biomedical literature. *Database (Oxford)* 2015:bav034. https://doi.org/10.1093/database/bav034.

M Cokol and R Rodriguez-Esteban. 2008. Visualizing evolution and impact of biomedical fields. *J Biomed Inform* 41(6):1050–1052. https://doi.org/10.1016/j.jbi.2008.05.002.

M Cokol, R Rodriguez-Esteban, and A Rzhetsky. 2007. A recipe for high impact. *Genome Biol* 8(5):406. https://doi.org/10.1186/gb-2007-8-5-406.

I Ezkurdia, D Juan, J Rodriguez, A Frankish, M Diekhans, J Harrow, J Vazquez, A Valencia, and M Tress. 2014. Multiple evidence strands suggest that there may be as few as 19 000 human protein-coding genes. *Hum Mol Genet* 23(22):5866–5878. https://doi.org/10.1093/hmg/ddu309.

L Furrer, S Clematide, H Marques, R Rodriguez-Esteban, M Romacker, and F Rinaldi. 2014. Collection-wide extraction of protein-protein interactions. *6th International Symposium on Semantic Mining in Biomedicine* pages 61–66. https://doi.org/10.5167/uzh-101472.

J Gobeill, E Pasche, D Vishnyakova, and P Ruch. 2013. Managing the data deluge: data-driven go category assignment improves while complexity of functional annotation increases. *Database (Oxford)* 2013:bat041. https://doi.org/10.1093/database/bat041.

J Hakenberg, M Gerner, M Haeussler, I Solt, C Plake, M Schroeder, G Gonzalez, G Nenadic, and C M Bergman. 2011. The GNAT library for local and remote gene mention normalization. *Bioinformatics* 27:2769–2771. https://doi.org/10.1093/bioinformatics/btr455.

J Hakenberg, C Plake, R Leaman, M Schroeder, and G Gonzalez. 2008. Inter-species normalization of gene mentions with GNAT. *Bioinformatics* 24:126–132. https://doi.org/10.1093/bioinformatics/btn299.

A Halevy, P Norvig, and F Pereira. 2009. The unreasonable effectiveness of data. *IEEE Intelligent Systems* 24(2):8–12. https://doi.org/10.1109/MIS.2009.36.

V Hatzivassiloglou, P A Dubou, and A Rzhetsky. 2001. Disambiguating proteins, genes, and rna in text: a machine learning approach. *Bioinformatics* 17 Suppl 1:S97–106. https://doi.org/10.1093/bioinformatics/17.suppl_1.S97.

L Hirschman, AA Morgan, and AS Yeh. 2002. Rutabaga by any other name: extracting biological names. *J Biomed Inform* 35(4):247–259. https://doi.org/10.1016/S1532-0464(03)00014-5.

S Jain, K R, T T Kuo, S Bhargava, G Lin, and C N Hsu. 2016. Weakly supervised learning of biomedical information extraction from curated data. *BMC Bioinformatics* 17 Suppl 1:1. https://doi.org/10.1186/s12859-015-0844-1.

R Kabiljo, A B Clegg, and A Shepherd. 2009. A realistic assessment of methods for extracting gene/protein interactions from free text. *BMC Bioinformatics* 10:233. https://doi.org/10.1186/1471-2105-10-233.

S Kaewphan, S Van Landeghem, T Ohta, Y Van de Peer, F Ginter, and S Pyysalo. 2016. Cell line name recognition in support of the identification of synthetic lethality in cancer from text. *Bioinformatics* 32(2):276–282. https://doi.org/10.1093/bioinformatics/btv570.

M Krauthammer and G Nenadic. 2004. Term identification in the biomedical literature. *J Biomed Inform* 37(6):512–526. https://doi.org/10.1016/j.jbi.2004.08.004.

R Leaman and G Gonzalez. 2008. BANNER: an executable survey of advances in biomedical named entity recognition. *Pac Symp Biocomput* pages 652–663. https://doi.org/10.1142/9789812776136_0062.

U Leser and J Hakenberg. 2005. What makes a gene name? Named entity recognition in the biomedical literature. *Brief Bioinform* 6(4):357–369. https://doi.org/10.1093/bib/6.4.357.

H Liu, ZZ Hu, M Torii, C Wu, and C Friedman. 2006. Quantitative assessment of dictionary-based protein named entity tagging. *J Am Med Inform Assoc* 13(5):497–507. https://doi.org/10.1197/jamia.M2085.

Z Lu, H Y Kao, C H Wei, M Huang, J Liu, C J Kuo, C N Hsu, R T Tsai, H J Dai, N Okazaki, H C Cho, M Gerner, I Solt, S Agarwal, F Liu, D Vishnyakova, P Ruch, M Romacker, F Rinaldi, S Bhattacharya, P Srinivasan, H Liu, M Torii, S Matos, D Campos, K Verspoor, K M Livingston, and W J Wilbur. 2011. The gene normalization task in BioCreative III. *BMC Bioinformatics* 12 Suppl 8:S2. https://doi.org/10.1186/1471-2105-12-S8-S2.

M Miwa, P Thompson, and S Ananiadou. 2012. Boosting automatic event extraction from the literature using domain adaptation and coreference resolution. *Bioinformatics* 28:1759–1765. https://doi.org/10.1093/bioinformatics/bts237.

A A Morgan, Z Lu, X Wang, A M Cohen, J Fluck, P Ruch, A Divoli, K Fundel, R Leaman, J Hakenberg, C Sun, H H Liu, R Torres, M Krauthammer, W W Lau, H Liu, C N Hsu, M Schuemie, K B Cohen, and L Hirschman. 2008. Overview of BioCreative II gene normalization. *Genome Biol* 9 Suppl 2:S3. https://doi.org/10.1186/gb-2008-9-s2-s3.

K Ravikumar, H Liu, J D Cohn, M E Wall, and K Verspoor. 2012. Literature mining of protein-residue associations with graph rules learned through distant supervision. *J Biomed Semantics* 3 Suppl 3:S2. https://doi.org/10.1186/2041-1480-3-S3-S2.

D Rebholz-Schuhmann, S Kafkas, J H Kim, C Li, A Jimeno Yepes, R Hoehndorf, R Backofen, and I Lewin. 2013. Evaluating gold standard corpora against gene/protein tagging solutions and lexical resources. *J Biomed Semantics* 4:28. https://doi.org/10.1186/2041-1480-4-28.

R Rodriguez-Esteban. 2016a. Additional knowledge-based analysis approaches. In W Loging, editor, *Bioinformatics and Computational Biology in Drug Discovery and Development*, Cambridge University Press, Cambridge, United Kingdom. https://doi.org/10.1017/CBO9780511989421.011.

R Rodriguez-Esteban. 2016b. Understanding human disease knowledge through text mining: What is text mining? In W Loging, editor, *Bioinformatics and Computational Biology in Drug Discovery and Development*, Cambridge University Press, Cambridge, United Kingdom. https://doi.org/10.1017/cbo9780511989421.004.

R Rodriguez-Esteban and W T Loging. 2013. Quantifying the complexity of medical research. *Bioinformatics* 29:2918–2924. https://doi.org/10.1093/bioinformatics/btt505.

M J Schuemie, B Mons, M Weeber, and J A Kors. 2007. Evaluation of techniques for increasing recall in a dictionary approach to gene and protein name identification. *J Biomed Inform* 40:316–324. https://doi.org/10.1016/j.jbi.2006.09.002.

J Wermter, K Tomanek, and U Hahn. 2009. High-performance gene name normalization with GeNo. *Bioinformatics* 25:815–821. https://doi.org/10.1093/bioinformatics/btp071.

M Zwick. 2015. Automated curation of gene name normalization results using the Konstanz information miner. *J Biomed Inform* 53:58–64. https://doi.org/10.1016/j.jbi.2014.08.016.

Towards Confidence Estimation for
Typed Protein-Protein Relation Extraction

Camilo Thorne and **Roman Klinger**
Institut für Maschinelle Sprachverarbeitung
University of Stuttgart, Stuttgart, Germany
{firstname.lastname}@ims.uni-stuttgart.de

Abstract

Systems which build on top of information extraction are typically challenged to extract knowledge that, while correct, is not yet well-known. We hypothesize that a good confidence measure for relational information has the property that such interesting information is found between information extracted with very high confidence and very low confidence. We discuss confidence estimation for the domain of biomedical protein-protein relation discovery in biomedical literature. As facts reported in papers take some time to be validated and recorded in biomedical databases, such task gives rise to large quantities of unknown but potentially true candidate relations. It is thus important to rank them based on supporting evidence rather than discard them. In this paper, we discuss this task and propose different approaches for confidence estimation and a pipeline to evaluate such methods. We show that the most straight-forward approach, a combination of different confidence measures from pipeline modules seems not to work well. We discuss this negative result and pinpoint potential future research directions.

1 Introduction

The ever increasing body of biomedical literature has motivated a growing interest over the past 20 years in natural language processing (NLP) and information extraction (IE) techniques to retrieve, organize and index the knowledge it contains (Rodriguez-Esteban, 2009; Subramaniam et al., 2003). It has also spurred a number of (shared) tasks and system competitions of which the best known are the BioNLP Shared Task[1] and the BioCreative challenge[2]. Relevant subtasks include named entity recognition (NER, Leaman and Gonzalez, 2008), entity linking and normalization to unique database identifiers (Zheng et al., 2014), event (EE, Björne and Salakoski, 2015) and relation extraction (RE, Tymoshenko et al., 2012; Airola et al., 2008; Choi, 2016). The overall goal is to identify biomedical entity mentions, disambiguate them w.r.t. biomedical databases and to identify mentioned biomedical relations and, crucially, *discover* new relations which are not available in structured resources yet.

When solving biomedical RE and IE tasks, the standard focus is to build systems that achieve high precision and recall at identifying *known relations* in gold standards or in biomedical databases and ontologies. This focus usually overlooks a key dimension for relation discovery: extraction *relevance* or *trust*. Indeed, when applied to new text in the form of, *e.g.*, recently published biomedical papers or papers from transversal domains such as bioinformatics, most discovered relations can arguably be expected to come up as "false", without being *per se* false – but unrecorded in gold standards. In other words, discovered relations fall under one of three categories: (1) plainly true relations (as per biomedical gold standards) (2) *interesting* relations that might be true or false. (3) plainly false relationships (as per biomedical gold standards). Our hypothesis is that a useful confidence measure estimates the quality of relations in this order, as we exemplify in Figure 1.

In such a scenario, rather than dismissing all such unknown (but interesting) relations, the goal is to return a ranking based on extraction *confi-*

---

[1] http://2016.bionlp-st.org/
[2] http://www.biocreative.org/

Proceedings of the Biomedical NLP Workshop associated with RANLP 2017, pages 55–63,
Varna, Bulgaria, 8 September 2017.

dence (Cullota and McCallum, 2004). Confidence typically refers to some kind of scoring – for instance a real number. This brings forth the problem of *confidence estimation*. While it is clear that the confidence of a relation extracted from biomedical text should be a function of the different sources of evidence on which it relies, it is unclear (Q1) how to define a global confidence estimator for biomedical relation extraction, and (Q2) how to evaluate it.

We hypothesize that relation discovery confidence scores rely on three main kinds of sources:

S1: The (aggregated) confidence scores of the individual modules of the RE pipeline.

S2: The internal graph structure of the discovered relations.

S3: Evidence gathered from external knowledge sources, such as textual evidence or knowledge retrieved or inferred from structured knowledge sources (biomedical ontologies and databases).

In this paper we outline a first attempt to answer questions (Q1) and (Q2) for the domain of protein-protein relations and events, focusing on approach S1. The main contributions of this paper are: (1) We build a distantly supervised RE pipeline based on BANNER (Leaman and Gonzalez, 2008) for NER, TEES (Björne and Salakoski, 2015) for EE and RE, and GNAT and Gnorm (Hackenberg et al., 2011; Wei et al., 2015) to link protein mentions to the STRING protein interaction database (von Mering et al., 2005), to distantly determine the truth and falsity of the discovered typed protein-protein relations. (2) We define confidence measures for each component of our pipeline and analyze their impact on relation prediction. (3) Finally, we propose and compare several global confidence estimators that aggregate over these scores.

2   Evidence Sources for Confidence

As said in the introduction, there are main sources of evidence for biomedical relation discovery and extraction, namely: prediction confidence (S1), graph analytics of the discovered relations (S2) and domain knowledge gathering (S3). We discuss these in the following.

S1   Modules (*e.g.*, NER or EE systems) in state-of-the-art systems are typically underpinned by

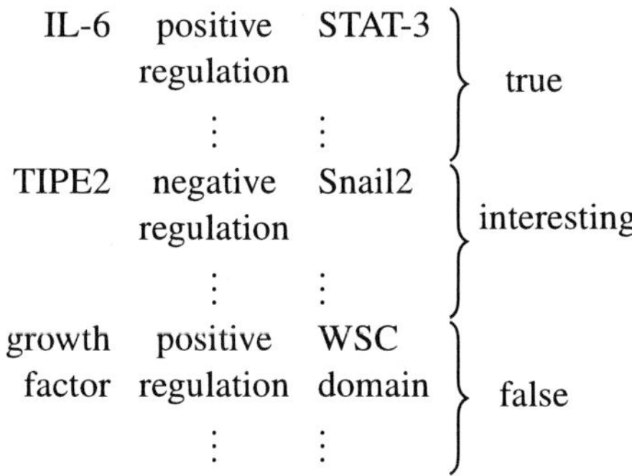

Figure 1: Discovered protein-protein (typed) relations. Notice how the bottom example is plainly false (it states a regulation among hormones), and the top one is plainly true (a known regulation). We assume the one in the middle to represent a more interesting result, as this fact is not in the STRING database; however, PubMed/MEDLINE abstract 28186089 mentions it ("(...) TIPE2 (...) downregulated (...) Snail2 (...)").

supervised classifiers that in addition to a prediction, return a probability (*e.g.*, logistic classifiers) or a so-called margin (*e.g.*, linear discriminant classifiers). We would expect interesting relations whose individual components (*e.g.*, entities and events) were identified with a higher score, to stand a higher chance of being true. To this end, one can employ confidence mixtures (Iversen et al., 2008; Dawid et al., 1995). Given k experts, each returning a confidence value $c_i \in \mathbb{R}$, $i = 1 \ldots k$, a *confidence aggregation* is a function $\varphi(\cdot)$ such that $c = \varphi(c_1, \ldots, c_k)$, where $c \in \mathbb{R}$ is the *global confidence* score. Global confidences thus aggregate partial confidences assigned to the partial tasks into which a complex task such as relation discovery can be broken down, to produce a global score. This method, the one actually described in this paper, can be seen as a baseline confidence estimator for biomedical RE.

S2   Graph-based confidence estimation techniques on the other hand rely on the graph-theoretical structure of extracted or discovered protein interactions and interaction networks. This makes sense because RE and EE systems (as the ones we rely on in this paper) actually return such graphs and interaction networks. In particular, one can leverage literature in biomedical and cross-domain *link prediction* (Lichtenwalter et al., 2010; Peng et al., 2017; And et al., 2003). Such techniques generally aim at predicting new edges (binary relations) in entity graphs via techniques such as similarity computation, weighted by properties such as the centrality or prominence of the connected entities – a measure that can be seen as a kind of confidence score. One can also exploit shortest path statistics among detected proteins (lengths of the paths, number of paths), as, intuitively the more relations between two proteins, the more likely that a specific relation holds.

S3   Last but not least, external knowledge sources can be used to, alone or in combination with the previous two methods, derive confidence estimators for biomedical relation discovery. Indeed, the STRING database itself describes a network of protein interactions, which can be combined with the interaction network built at discovery and extraction time to gather further, gold standard graph theoretical evidence for discovered interactions. Another possibility is to use techniques from the knowledge base population and enrichment communities such as Wick et al. (2013), reasoning over domain constraints and on whether discovered interactions satisfy or violate them (*e.g.*, the third example in Figure 1 is clearly false because its arguments are not proteins). Finally, one can also gather *textual evidence*, using techniques borrowed from cognitive systems such as IBM Watson (Murdock et al., 2012), exploiting PubMed/MEDLINE itself to derive lexical evidence.

3   Experiments

In this section, we describe our confidence estimation experiments for typed protein-protein interaction extraction and discovery. We refer to an *ordered* triple $rel = (p_1, r, p_2)$, where r is an event or relation type denoting a *directed* relation (*e.g.*, an expression, an inhibition) between proteins p_1 and p_2 as *typed interaction*. Note that this task is a subtask of event extraction (Kim et al., 2009) and an extension to protein-protein interaction detection (PPIs), where we want to predict if a protein pair (in any order) interacts in some way (Choi, 2016; Airola et al., 2008). Our whole pipeline is depicted in Figure 2.

3.1   Datasets

We used two main datasets in our experiments: Firstly, a large subset of MEDLINE from May 1992 to May 2017 (PMIDs 1376980 to 28211214). We

ignore languages other than English, and entries without abstract. We also disregarded abstracts that do not contain any mentions to protein or genes. This corpus consists of 40,911,675 tokens in 1,939,915 abstracts.

Secondly, to distantly evaluate discovered relations, we use the STRING database (von Mering et al., 2005), which describes protein-protein relations. STRING was built by integrating different databases (including the Gene Ontology (GO) and the Kyoto Encyclopedia of Genes and Genomes (KEGG)) and expert-curated text-mining-based information. STRING covers around 9.6 million protein entries and 1.3 billion interaction entries of 2031 unique organisms species. We focus on the subset of human proteins and their interactions. For network analysis, we use a Neo4j[3] graph database. From STRING, we use 20,458 unique genes/proteins with 6,013,567 unique typed interactions. They refer to 17,538 EntrezGene IDs.

3.2   Relation Extraction

To extract and discover relations in the MEDLINE subset, we rely on two well-known state-of-the-art systems for protein and gene detection and protein-protein event and relation extraction, namely BANNER and TEES (Leaman and Gonzalez, 2008; Björne and Salakoski, 2015).

BANNER is linear-chain conditional random field (CRF) NER system, that relies on an array of pre-trained models, dictionary and gold corpora for training and prediction. It uses the BIO format to spot the beginning (B) and constituent words (I) of a protein mention, and tokens that lie outside (O) mentions. For this paper, we use a gene detection model trained on the GNormPlus[4] gene gold corpus (Wei et al., 2015), which achieves 83 % F_1. Please note that we run BANNER separately from TEEs and realign the results in a separate step in the pipeline (see Figure 2).

TEES is a biomedical RE system, underpinned by a multiclass support vector machine (SVM). It relies on BANNER as a subcomponent to detect entities and on the BioNLP 2013 shared task data to estimate SVMs that detect (1) event triggering words and their GENIA event types: regulations, positive regulations, negative regulations, (de)phosphorilations (2) the arguments of

<hr>

[3]`https://neo4j.com/`

[4]We used this model for consistency with the GN systems that we describe below, which also use models trained over this corpus created for the BioCreative II GN shared task.

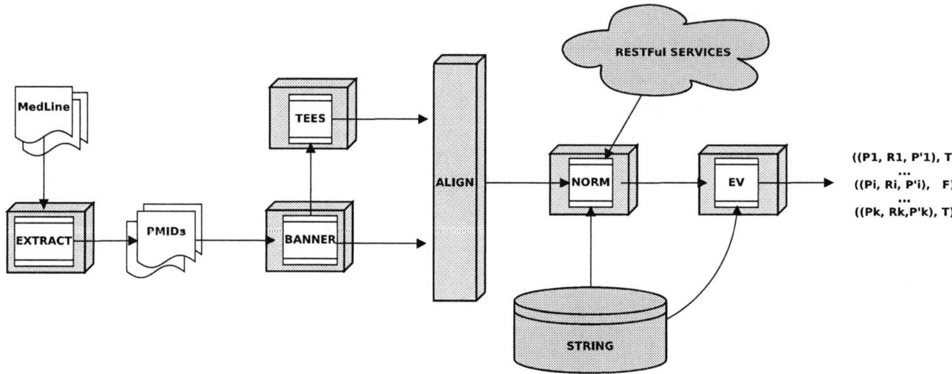

Figure 2: Overview of the relation extraction pipeline described in this paper (full pipeline).

	Unit	Count
	PMIDs	11773
	Relations	21169
Elements	Proteins	11726
	Events	864
	Causes	5694
	Themes	6032
Regul.	General	4425
	Positive	9830
	Negative	6484

Table 1: Event/Relation extraction statistics. "Events" refers to event trigger words, "Relations" refers to a relational structure connecting *typed* events to cause–theme protein pairs by TEES, as in Figure 3. All counts are unique counts.

the event or relation: its first argument (cause) and its second argument (theme). It can also detect complex event structures, event structures containing nested events, which we currently disregard. For each of its predictions, TEES returns a confidence value in the form of an SVM margin (distance of the trigger or protein to its separating hyperplane). TEES achieves 50.74 % F_1.

3.3 Relation Normalization

In order to verify if a candidate typed relation occurs in STRING, and to build (silver) standards for experimental analysis, we define a mapping from (M1) a protein mention p to a *canonical* form ($\mathrm{norm}(p)$), *i.e.*, STRING protein unique identifiers (UIDs), and (M2) a relation/event type r to a STRING interaction type ($\mathrm{ev}(r)$).

Protein matching Task (M1) is known in biomedical literature as the *protein normalization* task. It has been object of active research since the early 2000s, giving rise to the BioCreative Gene Normalization (GN) shared task. In this paper, we use two state-of-the-art GN systems, GNAT (Hackenberg et al., 2011), with a performance of 86.7 % F_1 and GNorm (Wei et al., 2015), with a performance of 86.4 % F_1. We denote this method by $\mathrm{gn_N}(p)$, for each GN system N and protein mention p. GNAT and GNorm normalize gene/protein mentions to EntrezGene UIDs, which cover a subset of STRING protein UIDs.

Therefore, in order to increase normalization recall, we resorted to a disambiguation-based method. We relied on a STRING RESTful webservice[5] that returns, given a protein mention p, a list of possible STRING canonical matches, together with a gloss (a small textual definition), to build a custom bag-of-words disambiguation method, that ranks candidates by computing the cosine similarity of the gloss and the sentence in which the mention occurs. We denote this method by $\mathrm{lk}(p)$.

This gave rise to a protein normalization method for protein mentions p summarized by:

$$\mathrm{norm_N}(p) = \begin{cases} \mathrm{gn_N}(p), & \text{if } \mathrm{gn_N}(p)\downarrow, \\ \mathrm{lk}(p), & \text{if } \mathrm{gn_N}(p)\uparrow, \mathrm{lk}(p)\downarrow, \quad (1) \\ \mathrm{NA}, & \text{if } \mathrm{gn_N}(p)\uparrow, \mathrm{lk}(p)\uparrow, \end{cases}$$

for $\mathrm{N} \in \{\mathrm{GNAT}, \mathrm{GNorm}\}$. By $\uparrow$ (resp. $\downarrow$) we mean that the method returns no canonical (resp. returns a canonical) STRING UID for mention p.

[5] https://string-db.org/cgi/help.pl?
&subpage=api

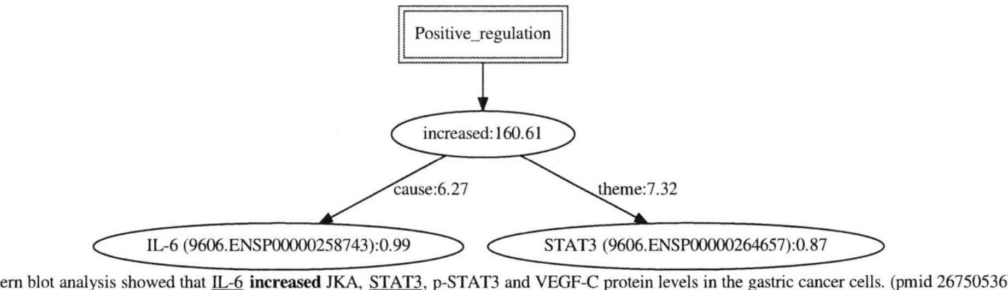

Figure 3: Protein-protein relational structure extracted by our pipeline for the first relation from Figure 1. The leave nodes represent the protein entities, labeled with their STRING UID and BANNER confidence. The internal node represents the event in which they participate as arguments, labeled with its TEES recognition confidence. The labels on its outgoing edges represent cause–theme TEES labeling of its protein arguments, and its TEES confidence. Finally, the root represents the predicted event type.

Note that GNAT and GNorm were tuned for distinct, though related GN subtasks, namely human GN and cross species GN, and can produce different results. If no normalization method returns a STRING UID, we consider the canonical protein for mention p undefined (NA).

Event type matching To deal with (M2), we relied on the other hand on a simple rule-based method, that maps the three GENIA event types returned by TEES GENIA event types to typed and directed interactions in STRING: protein inhibitions, activations and expressions. As a GENIA event type r may correspond to more than one STRING interaction, we map them to *sets* of interactions with

$$\mathrm{ev}(r) = \begin{cases} \{\text{inhibitits}\}, & \text{if } r = R^-, \\ \{\text{expresses,activates}\}, & \text{if } r = R^+, \\ \{\text{expresses,activates}\} \\ \cup\{\text{inhibits}\} & , \text{ if } r = R. \end{cases} \quad (2)$$

In other words, a TEES relation type r (a regulation R, a negative regulation R^-, or a positive regulation R^+) will be mapped to (sets of) STRING protein inhibitions, expressions and activations.

Relation matching For $N \in \{\text{GNAT, GNorm}\}$, we determine a positive match for a candidate relation (triple) (p_1, r, p_2) if for at least a value $t \in \mathrm{ev}(r)$ the triple $(\mathrm{norm}_N(p_1), t, \mathrm{norm}_N(p_2))$ occurs in the STRING database, negative otherwise. If $\mathrm{norm}_N(p_i)$, for $i \in \{1, 2\}$, returns no canonical STRING UID, we discard the candidate altogether. As GNAT and GNorm produce different

normalizer	norm. relations	positive
GNAT*	11723	973
GNormPlus*	8639	544

Table 2: Silver standards obtained with our normalization methods. By the asterisk we mean the GN system plus our backoffs. By "norm. relations" we mean the number of relational structures for which protein pairs and event types could be normalized to STRING interaction types and protein UIDs and by "positive" to those that actually match interactions in STRING.

results, rather than aggregating results, we generated two separate silver standards, summarized by Table 2. Both cover around 1/2 of the original dataset of candidates and both are skewed towards negative matches.

3.4 Confidence Estimation

In this subsection we describe our global confidence estimation models. These models aggregate confidence values returned by the key components of our pipeline, namely, BANNER and TEES for proteins, and event/event types, as shown in Figure 3.

Component-wise confidence For every RE candidate triple $rel = (p_1, r, p_2)$ we compute the following confidence values:

Entity-level (marginal) confidence (BANNER): we return the so-called *product gamma probability* (Cullota and McCallum, 2004) of protein theme (resp. cause) mentions p starting at position

t in an abstract with BIO labels $(s_t, \ldots, s_{t+k})$, defined by:

$$\mathrm{cf}_{\gamma_t}(p) = \prod_{i=t}^{k} \gamma_i(s_i) \qquad (3)$$

(resp., $\mathrm{cf}_{\gamma_c}(p)$ for cause mentions) where $\gamma_i(s_i) = \alpha_i(s_i) \cdot \beta_i(s_i)/P(w_0, \ldots, w_i; \Lambda)$ is the normalized product of the forward and backward Viterbi lattice probabilities of label $s_i \in \{\mathrm{B}, \mathrm{I}\}$ at position i, computed from BANNER's underlying CRF model Λ, and w_i is a word token. This measure basically characterizes the likelihood that a given span of MEDLINE tokens is indeed a protein.

From TEES, we use event-level confidence based on the margins in the SVM, namely $\mathrm{cf}_{ev}(r)$, $\mathrm{cf}_c(p_1)$ and $\mathrm{cf}_t(p_2)$ for event (type), cause, and theme predictions.

In summary, we use *five* component-wise confidence features for a relation triple $rel = (p_1, r, p_2)$, namely, the BANNER product gamma probability of theme proteins, the TEES margin value for theme proteins, the BANNER product gamma probability of cause proteins, the TEES margin value for cause proteins, and the TEES margin value for events/event types.

Confidence aggregation Different confidence sources will have a different impact on their global aggregate (Iversen et al., 2008; Dawid et al., 1995). Such impact can be quantified as a *weight*, set *a priori* or *a posteriori* by training a classifier over gold (or silver) data and plugging into the aggregates the inferred weights (Liu et al., 2012). For the experiments described in this paper we chose the latter, and trained a logistic classifier over our silver MEDLINE datasets (see the next subsection for a detailed description), and used its coefficients $\vec{\theta}$ to compute the weights $\vec{we}$ by (1) measuring their impact on classification deviance (see Table 5), and (2) normalizing the values to a number between 0 and 1. We propose two fundamentally different methods to aggregate the separate confidence values to one measure for a triple rel.

The first method assumes that global confidence is a linear combination of component-wise confidences for a relation rel (cf_m, with $m \in \{\gamma_t, \gamma_c, ev, c(p_1), t(p_2)\}$), namely, their *(weighted) average*:

$$\mathrm{cf}_{avg} = \frac{1}{5} \cdot \sum_m we_m \cdot \mathrm{cf}_m \qquad (4)$$

The second method assumes that each component-wise confidence is totally independent of each other (and hence independence for each pipeline prediction), and defines global confidence as a *(weighted) product*:

$$\mathrm{cf}_{prod} = \prod_m we_m \cdot \mathrm{cf}_m \qquad (5)$$

We considered also unweighted versions of the confidence aggregators, by considering unit weights $\vec{we} = (1, 1, 1, 1, 1)^T$, that assign the same importance to all component-wise confidences.

Evaluation To evaluate our approach, we relied on a number of different strategies and combinations thereof. In particular, we split our two silver GNAT and GNORM datasets $\mathcal{S}_N$, into two disjoint train $\mathcal{T}_N$ and test $\mathcal{E}_N$ subsets. Given how unbalanced our data is, we, in addition, resampled the training sets by (1) oversampling positive matches, and (2) undersampling negative matches until we obtained two balanced training sets $\mathcal{S}_{GNAT}$ and $\mathcal{S}_{GNorm}$ each of 2000 relations. For testing, we kept a set of 1000 unresampled relations each.

To learn the weights $\vec{\theta}$ of the confidence aggregation models and hence of component-wise confidences, we trained a logistic classifier over each of our silver standards:

$$P(t'\!=\!1|\vec{c}) = (1 + \exp(-\sum_m \theta_m \cdot \mathrm{cf}_m))^{-1} \quad (6)$$

where $t' = 1$ if *normalized* triple rel is in STRING, $m \in \{\gamma_t, \gamma_c, ev, c(p_1), t(p_2)\}$ and $\vec{c} = (\mathrm{cf}_{\gamma_t}, \mathrm{cf}_{\gamma_c}, \mathrm{cf}_{ev}, \mathrm{cf}_{c(p_1)}, \mathrm{cf}_{t(p_2)})^T$. The parameters $\vec{\theta}$ were learned by maximizing the likelihood $\mathcal{L}(\vec{\theta}; \mathcal{T}_N) = \prod_j \pi(\vec{c}^{(j)}; \vec{\theta})^{r'^{(j)}} \cdot (1 -$

train dataset $\mathcal{T}$	test dataset $\mathcal{E}$	F_1
gnat_train	test_gnat	0.688
gnorm_train	test_gnat	0.658
gnat_train	**test_gnorm**	**0.766**
gnorm_train	test_gnorm	0.733

Table 3: Evaluation of logistic models over the different possible train/test combinations of our various silver standards. In bold, the combination with the best performance. We used the best model (gnat_train) to derive the logistic model (Equation 6) and the weights used in the weighted confidence aggregation models.

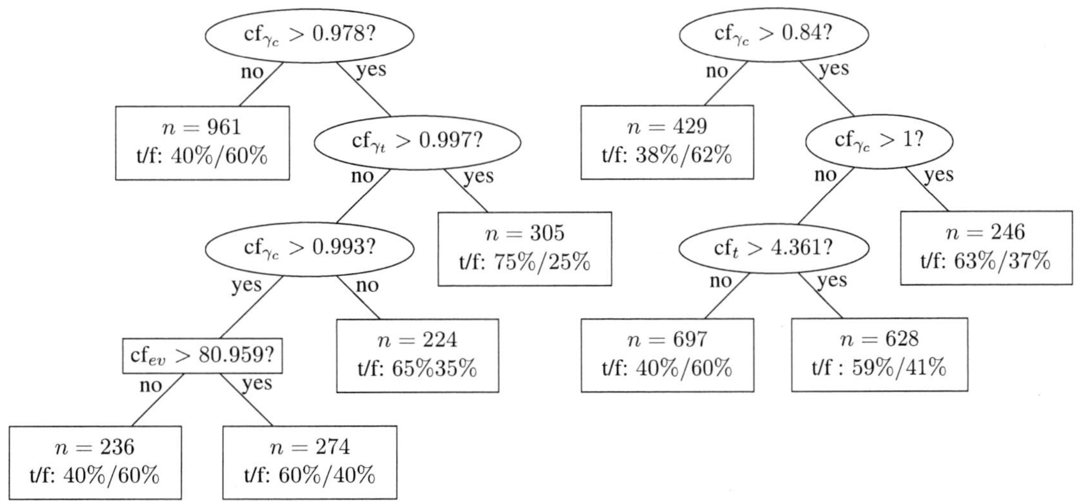

Figure 4: Right: J48 decision tree for the GNAT silver standard (training set). Left: J48 decision tree for the GNorm silver standard (training set). Nodes correspond to the component-wise confidence features defined in Section 3.4. The higher a component-wise confidence, the higher its information gain. Notice how, in general, we observe a higher gain for TEES confidence scores, plus some contribution coming from the BANNER confidence of theme proteins. We used for both models a pruning setup whereby we imposed each tree leave to contain at least 150 relations. In the visualization, the leaves describe also the distribution of positive (t) and negative (f) matches for each bin, and their size n (triples per bin).

estimator	Kendall τ	p-value
cf_{prod} (unweig.)	0.041	0.127
cf_{prod}	0.041	0.127
cf_{avg} (unweig.)	0.032	0.210
cf_{avg}	**0.050**	**0.056**

Table 4: Correlation-based evaluation of the confidence aggregation models. In bold, the model with the highest τ value. No test was statistically significant (although one came close to $p = 0.05$). In all cases, this indicates absence of correlation with linking judgments. Unweighted models were obtained by considering uniform weights (viz., $\vec{we} = (1, 1, 1, 1, 1)^T$).

feature	deviance	p-value
cf_{ev}	6.200	0.013
cf_t	**17.803**	**$2.451 \cdot 10^{-05}$**
cf_c	4.858	0.028
cf_{γ_t}	2.370	0.124
cf_{γ_c}	**22.667**	**$1.926 \cdot 10^{-06}$**

Table 5: ANOVA/Analysis of deviance table for the best logistic model from Table 3 (χ^2-test). In bold, the features with greater impact, both statistically significant with $p < 0.01$.

$\pi(\vec{c}^{(j)}; \vec{\theta}))^{1-r'^{(j)}}$ via iterative weighted least squares. We tested each of the two ensuing logistic models over each of the two test datasets, and chose the model with the highest F_1, as seen in Table 3.

The confidence estimation models themselves were evaluated following a methodology proposed by Cullota and McCallum (2004) to evaluate entity-level confidence measures: measure the correlation between matching judgments and relation confidence. Ideally, one would expect a bias in confidence towards positive matches. In this paper, we considered Kendall's τ correlation, which is rank-sensitive and robust to ties.

Last, but not least, we used the logistic model and the balanced datasets to conduct an exploratory analysis on the component-wise confidence themselves, to understand which, from all of our pipeline's components has a bigger impact on global confidence estimation. To this end we relied on two separate methodologies: On the one hand, we conducted an analysis of variance/deviance[6] over the (optimal) logistic model's features. On the other hand, we inferred two decision trees over our two training sets. Decision trees rank component-wise confidences cf_m, w.r.t. *information gain*. We used the J48 decision tree classifier[7], that discretizes continuous variables.

4 Results and Discussion

The results of our confidence aggregation experiments are summarized by Tables 3–5 and Figure 4.

As Table 3 shows, the best logistic model was obtained over the GNAT training dataset. Interestingly, the best result arose from cross-testing, when be tested it of the GNorm dataset test corpus. We conjecture that this might be due to a slightly better generalization capacity of the GNAT normalizer, as opposed to GNorm.

Regarding our confidence estimation models however, as Table 4 shows, our analysis returned no observable correlation (all τ values are close to zero), but without reaching statistical significance. Furthermore, of all estimators, the best (albeit by a very small margin) estimator was average, weighted confidence. We interpret this negative result to mean that aggregating confidences alone, disregarding: (1) the performance and/or confidence of the different normalizations methods (2) the structural properties of discovered relations, and (3) additional evidence gathered from external sources is simply not enough to define meaningful confidence estimators.

Finally, as shown by Table 5 and Figure 4 both the ANOVA and decision-tree/information gain analysis point out that the most informative features were the BANNER and TEES confidences for the arguments – the theme (2nd argument) and the cause (1st argument) – of protein-protein relational structures. Interestingly event (TEES) confidences do not seem to play a major role. This however seems consistent with the fact that TEES models are optimized for recognizing theme-event and cause-event pairs (by leveraging on the dependency parse tree of the sentence), a harder task than that of event recognition.

It suggests that while the aggregation of component-wise confidences is not a good global confidence estimator, including them as features of a more complex model encompassing a wider array of evidence sources might still be useful. It also suggests that normalization confidence – the last step in the pipeline – should be taken into account as the confidence values coming from protein recognition have the most impact.

5 Conclusions & Future Work

In this paper we have proposed a confidence estimation methodology for biomedical protein-protein typed interaction discovery from PubMed/MEDLINE abstracts. Measuring confidence or trust is important because in this setting not all false positives – interactions that are not known to occur in biomedical databases – may be necessarily false. This sorting by confidence should satisfy key criteria, namely that true matches should be scored high, clearly false matches low, and "interesting" relations somewhere in between.

To do so, we have proposed a pipeline that builds upon state-of-the-art protein NER, protein-protein EE and RE and GN systems, to discover and distantly evaluate against the STRING database protein-protein typed interactions. Then, we have described a number of baseline confidence estimation techniques that aggregate the confidence prediction scores of the pipeline's components.

Our experiments and correlation analysis show that, while the prediction confidence of modules in later stages of the pipeline seems to influence more positive decisions, confidence aggregation is not enough to define estimation models satisfying the criteria mentioned. We conjecture that this is due to the fact that prediction confidence alone does not provide sufficient evidence to rank relations. Also, in this work, the confidence of normalization was not fully addressed. As further work we plan to focus on more complex evidence gathering methods.

Aknowledgements We thank Jörg Hackenberg for his help running and integrating GNAT into our pipeline, and Philippe Thomas for his comments on our event mappings. This work was supported by a grant from the Ministry of Science, Research and Arts of Baden-Württemberg to Roman Klinger.

---

[6]Logistic models are known in statistical literature as *generalized linear models*; in such cases rather than analyzing error variance as for linear models, one analyzes *deviance*, viz., prediction error.

[7]Weka (http://www.cs.waikato.ac.nz/ml/weka/) for our logistic and J48 models.

References

Antti Airola, Sampo Pyysalo, Jari Björne, Tapio Pahikkala, Filip Ginter, and Tapio Salakoski. 2008. All-paths graph kernel for protein-protein interaction extraction with evaluation of cross-corpus learning. *BMC Bioinformatics* 9(Suppl 11):S2.

Dennis Wilkinson And, Dennis Wilkinson, and Bernardo A. Huberman. 2003. A method for finding communities of related genes. In *Proceedings of the National Academy of Sciences of the United States of America*. pages 5241–5248.

Jari Björne and Taio Salakoski. 2015. Tees 2.2: Biomedical event extraction for diverse corpora. *BMC Bioinformatics* 16(16):S4.

Sung-Pil Choi. 2016. Extraction of proteinprotein interactions (PPIs) from the literature by deep convolutional neural networks with various feature embeddings. *Journal of Information Science* .

Aron Cullota and Andrew McCallum. 2004. Confidence estimation for information extraction. In *Proceedings of the 2004 Conference of the North American Chapter of the Association for Computational Linguistics*. HLT-NAACL '04, pages 109–112.

A. Dawid, M. DeGroot, J. Mortera, Roger Cooke, S. French, C. Genest, M. Schervish, D. Lindley, K. McConway, and R. Winkler. 1995. Coherent combination of experts' opinions. *TEST: An Official Journal of the Spanish Society of Statistics and Operations Research* 4(2):263–313.

Jög Hackenberg, Marin Gerner, Maximilian Haeussler, Illés Solt, Conrad Plake, Martin Schroder, Graciela Gonzalez, Goran Nenadic, and Casey M. Bergman. 2011. The GNAT library for local and remote gene mention normalization. *Bionformatics* 27(19):2769–2771.

Edwin S Iversen, Giovanni Parmigiani, and Sining Chen. 2008. Multiple model evaluation absent the gold standard through model combination. *Journal of the American Statistical Association* 103(483):897–909.

Jin-Dong Kim, Tomoko Ohta, Sampo Pyysalo, Yoshinobu Kano, and Jun'ichi Tsujii. 2009. Overview of BioNLP '09 shared task on event extraction. In *Proceedings of the Workshop on Current Trends in Biomedical Natural Language Processing: Shared Task*. BioNLP '09, pages 1–9.

Robert Leaman and Graciela Gonzalez. 2008. BANNER: An executable survey of advances in biomedical named entity recognition. In *Proceedings of the 2008 Pacific Symposium on Biocomputing*. PSB '08, pages 652–63.

Ryan N. Lichtenwalter, Jake T. Lussier, and Nitesh V. Chawla. 2010. New perspectives and methods in link prediction. In *Proceedings of the 16th ACM SIGKDD International Conference on Knowledge Discovery and Data Mining*. KDD '10, pages 243–252.

X. Liu, Amol Ghorpade, Y. L. Tu, and W. J. Zhang. 2012. A novel approach to probability distribution aggregation. *Information Science* 188:269–275.

J. William Murdock, James Fan, Adam Lally, Hideki Shima, and Branimir Boguraev. 2012. Textual evidence gathering and analysis. *IBM Journal of Research and Development* 56(3):8.

Jiajie Peng, Kun Bai, Xuequn Shang, Guohua Wang, Hansheng Xue, Shuilin Jin, Liang Cheng, Yadong Wang, and Jin Chen. 2017. Predicting disease-related genes using integrated biomedical networks. *BMC genomics* 18(1):1043.

Raul Rodriguez-Esteban. 2009. Biomedical Text Mining and Its Applications. *PLoS Comput Biol* 5(12).

L. Venkata Subramaniam, Sougata Mukherjea, Pankaj Kankar, Biplav Srivastava, Vishal S. Batra, Pasumarti V. Kamesam, and Ravi Kothari. 2003. Information extraction from biomedical literature: Methodology, evaluation and an application. In *Proceedings of the Twelfth International Conference on Information and Knowledge Management*. CIKM '03, pages 410–417.

Kateryna Tymoshenko, Swapna Somasundaran, Vinodkumar Prabhakaran, and Vinay Shet. 2012. Relation mining in the biomedical domain using entity-level semantics. In *Proceedings of the 20th European Conference on Artificial Intelligence*. ECAI '12, pages 780–785.

Christian von Mering, Lars J. Jensen, Berend Snel, Sean D. Hooper, Markus Krupp, Mathidel Foglierini, Nelly Jouffre, Martijn A. Huynen, and Peer Bork. 2005. STRING: Known and predicted protein-protein associations, integrated and transferred across organisms. *Nucleic Acids Research* 33(Suppl. 1):D433–D437.

Chih-Husuan Wei, Hung-Yu Kao, and Zhiyoung Lu. 2015. GNomrPlus: An integrative approach for tagging genes, gene families, and protein domains. *BioMed Research International* 2015(2015):ID 918710.

Michael L. Wick, Sameer Singh, Ari Kobren, and Andrew McCallum. 2013. Assessing confidence of knowledge base content with an experimental study in entity resolution. In *Proceedings of the 2013 Workshop on Automated Knowledge Base Construction*. AKBC@CIKM '13, pages 13–18.

Jin Guang Zheng, Daniel Howsmon, Boliang Zhang, Juergen Hahn, Deborah McGuinness, James Hendler, and Heng Ji. 2014. Entity linking for biomedical literature. In *Proceedings of the ACM 8th International Workshop on Data and Text Mining in Bioinformatics*. DTMBIO '14, pages 3–4.

Identification of Risk Factors in Clinical Texts through Association Rules

Svetla Boytcheva[1]   Ivelina Nikolova[1]   Galia Angelova[1]   Zhivko Angelov[2]

[1] Institute of Information and Communication Technologies,
Bulgarian Academy of Sciences
[2] Adiss Lab Ltd., Sofia, Bulgaria
svetla.boytcheva@gmail.com {iva,galia}@lml.bas.bg,
angelov@adiss-bg.com

Abstract

We describe a method which extracts Association Rules from texts in order to recognise verbalisations of risk factors. Usually some basic vocabulary about risk factors is known but medical conditions are expressed in clinical narratives with much higher variety. We propose an approach for data-driven learning of specialised medical vocabulary which, once collected, enables early alerting of potentially affected patients. The method is illustrated by experimens with clinical records of patients with Chronic Obstructive Pulmonary Disease (COPD) and comorbidity of CORD, Diabetes Melitus and Schizophrenia. Our input data come from the Bulgarian Diabetic Register, which is built using a pseudonymised collection of outpatient records for about 500,000 diabetic patients. The generated Association Rules for CORD are analysed in the context of demographic, gender, and age information. Valuable anounts of meaningful words, signalling risk factors, are discovered with high precision and confidence.

1   Introduction

Chronic diseases like Chronic Obstructive Pulmonary Disease (COPD) and Diabetes Mellitus are long-lasting disorders with effects that come with time. They are the result of a combination of genetic, physiological, environmental and behaviours factors, and kill over 40 million people each year, equivalent to 70% of all deaths globally[1]. Prevention is focused on reducing the risk factors associated with these diseases. Therefore, establishing the risk rates and early recognition of potential danger will help to decrease the role of the common modifiable risk factors. In the age of big data and given the growing amount of patient-related texts, we believe that Data Mining and Text Mining are key technologies which might help by providing discovery of hidden interdependencies among words (lexical expressions of indicators and assessment of risks) in patient records.

In this paper we demonstrate how automatic analysis of clinical narratives in Bulgarian language allows to identify verbal expressions of risks for patients. Our input data come from the Bulgarian Diabetic Register, which is built using a pseudonymised collection of outpatient records for about 500,000 diabetic patients treated in the period 2010-2016 (Tcharaktchiev et al., 2015). Together with the structured information, the outpatient records contain free texts discussing the patient case history, status, risk factors, treatment etc. Our tools process both structured data and free text of outpatient records in order to extract Association Rules for COPD risk factors. Since Diabetes Melitus and Schizophrenia are also closely related, we study their comorbidity and the risk factors for COPD in patients with Diabetes Melitus and Schizophrenia. By applying unsupervised Data Mining techniques we try to overcome the lack of linguistic and ontological resources that can support successful NLP analysis of clinical narratives in Bulgarian. Thus we demonstrate how new lexical resources can be generated, to be used for better analysis of clinical texts.

The paper is structured as follows. Section 2 overviews related work with focus on the technological solutions. Section 3 presents the method we use, section 4 – the experiments and results. Section 5 contains the conclusion and discusses future work.

---

[1] World Health Organisation (WHO) factsheets: http://www.who.int/mediacentre/factsheets/fs355/en/

Proceedings of the Biomedical NLP Workshop associated with RANLP 2017, pages 64–72,
Varna, Bulgaria, 8 September 2017.

2   Related Work

Many advanced approaches apply Natural Language Processing (NLP) as a first step in mining entities from free texts and use the latter as input to subsequent biomedical research or decision making tasks. Incorporating NLP has advantages: it systematically links several terms to a concept using databases that standardise health terminologies; avoids manual work for searching term variations; increases the number of patients in the considered cohorts and thus increases the sensitivity of the recognition (Liao et al., 2015). A recent review lists 71 clinical NLP systems, which process free text and generate structured output, in order to address a wide variety of important clinical and research tasks (Kreimeyer et al., 2017). Significant progress has been made in algorithm development and resource construction since 2000 (Luo et al., 2017). Open challenges remain e.g. extraction of temporal information, normalisation of concepts to standard terminologies, interpretation etc. Despite the limitations the conclusion is that today NLP engines are powerful components ready for integration in medical text processing and – due to expected improvements in the near future, e.g. more accurate mappings of terms to medical concepts – the importance of NLP as a valuable supporting technology will grow (Liao et al., 2015). Here we briefly discuss major text analysis technologies that are applied in biomedical domain.

Data mining (DM) is actively used in the field since the middle of 1990's. It employs explorative algorithms to identify meaningful data patterns with acceptable computational efficiency and uncover new biomedical and healthcare knowledge for clinical and administrative decision making. Furthermore it can generate testable evidence-based medical hypotheses from large experimental data, clinical databases, and/or biomedical literature. Today DM is applied for a variety of tasks operating on biomedical entities extracted from free texts. For instance (Luo et al., 2017) states that NLP is a useful tool for extracting information related to adverse drug events (ADE) and pharmaceutical products from electronic health record (EHR) narratives. Since 2012, DM enables successful automation of the ADE discovery so the "NLP-based ADE detection" (as the authors call it) can be soon integrated in practical systems. Moreover, the DM capacity for treatment of heterogeneous data sources is increasingly adopted.

(Stubbs et al., 2015) present an overview of the 2014 i2b2/UTHealth NLP shared task focused on identifying medical risk factors related to Coronary Artery Disease (CAD) in the narratives of longitudinal medical records of diabetic patients. Twenty teams participated in this track, and submitted 49 system runs for evaluation. The most successful system used a combination of external lexicons, hand-written rules and Support Vector Machines (a machine learning method). Other machine learning techniques n use were Conditional Random Fields and ensembles of classifiers (CRF, Naïve Bayes, and Maximum Entropy). With six of the top 10 teams achieving F1 scores over 0.90, and all 10 achieving F1 scores over 0.87, the authors conclude that identification of risk factors and their progression over time is within the reach of present automated systems. These examples show that today DM is a key technology for the successful NLP-based medical applications.

Text mining (TM) aims at the delivering of meaningful information from texts, e.g. structuring text units into entities and relationships among them, via NLP applications for shallow analysis. A widely used system of this type is the open-source NLP tool for information extraction from EHR cTAKES (clinical Text Analysis and Knowledge Extraction System)[2]. Another open source system is HITEx (Health Information Text Extraction) which extracts some variables of interest from narrative text (Goryachev et al., 2006). We mention here two more examples how text mining delivers useful information about risk factors and adverse drug events. In (Jonnagaddala et al., 2015) the authors present a system that discovers in free text EHRs information about age, gender, total cholesterol (or low-density lipoproteins cholesterol LDL-C), high-density lipoproteins cholesterol (HDL-C), blood pressure, diabetes history and smoking history for a cohort of 164 diabetic patients. After that the Framingham risk score is calculated to predict the coronary artery disease (CAD) for these patients. The performance of the text extraction system is reliable, however missing data remain a challenging issue. Over 40% of patients in the final cohort are at high risk of CAD and over 50% of the population fitted in the moderate category. The main limitation was the lack of a systematic evaluation of the developed text mining system. In (Harpaz

---

et al., 2014) the authors state that TM is sufficiently mature to be applied for the extraction of useful information concerning ADEs from multiple textual sources. Currently such information is collected by manual expert analysis of clinical trial notes and spontaneous reports, and the review of biomedical literature; but progress depends on a comprehensive approach that examines a diverse set of potentially complementing data sources including EHRs. Posting in social media are another source of information about ADEs: 2% of patients and 6% of caregivers share their experiences online, and 18% of all internet users, 31% of all patients with chronic conditions, and 38% of caregivers look at online drug reviews[3]. Despite the challenges, a large body of research has demonstrated that the existing TM tools are capable to extract useful safety-related information from the aforementioned textual sources.

NER and **rule-based approaches** evolved during the last decades from research prototypes to reliable NLP technologies. Mature (and constantly evolving) systems appeared for processing English clinical texts, e.g. KnowledgeMap Concept Identifier which processes clinical notes and returns CUIs (Concept Unique Identifiers) for the recognized UMLS terms (Denny et al., 2003) as well as NegEx, a tool for identification and interpretation of negation in English texts (Chapman et al., 2001), (Gindl, 2006). Identification of temporal events is a hot topic in biomedical NLP. In (Chang et al., 2015) it is proposed to recognise first all temporal expressions and then, after building a temporal model of the context, to assign the corresponding time attributes for all recognised concepts with respect to the creation time of the records. Disease mentions are identified after that, along with their corresponding risk factors and medications. (Chang et al., 2015) shows the progress in processing named entities which represent temporal information. Recently, with the DM development, classical rule-based systems like NegEx can be outperformed by statistical methods (Uzuner et al., 2009); on the other hand the rule-based methods prove to be good in the production of annotated resources and when writing rules that emulate the knowledge of a domain expert (e.g. in ADE discovery).

3 Methods

Our approach (Fig. 1) has five main phases: *(i)* Structured information processing of the ORs in the repository; *(ii)* Risk Factors Association Rules generation from the training set; *(iii)* Preprocessing of the test sets; *(iv)* Risk Factors Association Rules matching on the test sets; *(v)* Structured information processing of the patients in risk.

3.1 Structured Data Analysis Methods

The Diabetes Register contains pseudoanonymous Outpatient Records (OR) in XML format. Most data necessary for the health management are structured in fields with XML tags which present the Patient ID, the code of doctors' medical specialty, region of practice, Date/Time and ID of the OR. Several free-text fields contain important explanations about the patient: "Anamnesis", "Status", "Clinical examinations" and "Therapy". There are also several XML tags for the main diagnose and additional diagnoses with their codes according to the International Classification of Diseases, 10th Revision (ICD-10)[4]. Each OR contains a main diagnosis with ICD-10 code and ICD-10 codes of up to 4 additional disorders, i.e. in total from 1 to 5 ICD-10 codes.

The study of disorder comorbidities plays an important role in detection and prevention of patients at risk. Chronic diseases constitute a major cause of mortality according to the World Health Organization (WHO) reports and their study is of higher importance for healthcare. For discovering frequent patterns of chronic diseases we use retrospective analysis of population data, by filtering events with common properties and similar significance. One of the major approaches to pattern search is frequent pattern mining (FPM) viewing the events (objects) as unordered sets. This preliminary work was done over outpatient records (ORs) of patients with primary diagnose Diabetes Melitus Type 2 (ICD-10 code E11) *(withdrawn Self-reference)*. We extracted relatively high number of frequent patterns containing different mental disorders – ICD-10 codes F00-F99. This result motivated us to process collection for patients with Schizophrenia (ICD-10 code F20). The study collection *SD* of patients who suffer from both Schizophrenia and Diabetes Melitus Type 2 was

---

[3]Pew Research Center, The Social Life of Health Information, 2011: http://www.pewinternet.org/2011/05/12/the-social-life-of-health-information-2011/

[4] International Classification of Diseases and Related Health Problems 10th Revision. http://apps.who.int/classifications/icd10/browse/2015/en

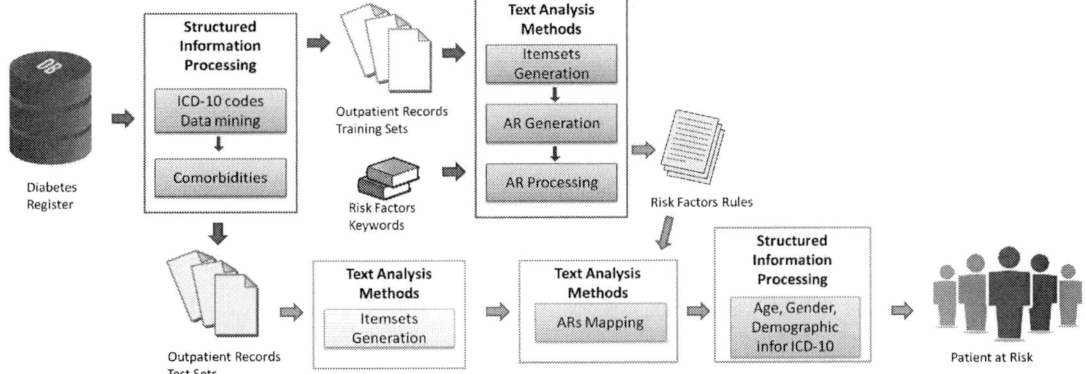

Figure 1: Identification of patients at risk

automatically extracted from the Diabetes Register and contains all ORs for these patient in the period 2012-2014 - approx. 200,000 ORs for 4,080 patients.

Let H be a chronic disease and there exist a frequent itemset F of chronic diseases such that $\{H, E11, F20\} \subseteq F$. The study collection SD is split into two subsets SH and ST. The collection SH contains ORs of all patients in SD that also have diagnosis H. We will call the set SH a training set. The set ST is formed as $ST = SD - SH$. We will call the set ST a test set.

3.2 Risk Factors Association Rules Generation

Text Analysis has three main phases: *Itemsets Generation* which converts the text documents into itemsets, *Association Rules Generation* based on frequent pattern mining (FPM) techniques and elicitation of ARs, and *Risk Factors Association Rules Filtering* that filters rules by using keywords (Fig. 2).

The system processes input texts in unicode format and is language independent in principle (stemming and stopword filtering can be replaced with modules for another language).

3.2.1 Itemsets Generation

Let SH be the training set. We extract for each OR its parts in XML tags for Anamnesis (Patient History) and Status and form separate collections of ORs Anamnesis texts only - SHa, and ORs Status texts only - SHh correspondingly. We process separately the collections SHa and SHh.

Let S be one collection. Each text in S is turned to a sequence of word stems in their original order, using blank spaces and punctuation delimiters as tokenization separators. Stop words and numbers may be essential for some patterns so they are preserved and generalised - replaced by the constants STOP and NUM correspondingly. After this step the punctuation is eliminated. Then we use hashing and substitute each word with an unique number. In addition some compression and sorting is applied. This is necessary to speed up the frequent patterns mining process.

The vocabulary used in all documents of S will be called *items* $W = \{w_1, w_2, ..., w_n\}$. For the collection S we extract the set of all different documents $P = \{p_1, p_2, ..., p_N\}$, where $p_i \subseteq W$. This set corresponds to transactions; the associated unique transaction identifiers (*tids*) shall be called **pids** (patient identifiers). Each patient interaction with a doctor is viewed as a single document in P.

3.2.2 Association Rules Generation

The ORs are written in telegraphic style with phrases rather than full sentences. Usually the ORs list attribute-value (*A-V*) pairs - anatomical organ/system and its status/condition. Attribute names contain phrases and abbreviations in Cyrillic and Latin. Values can be long descriptions in case of status complications. The order of *A-V* pairs can vary and parts of the value descriptions can surround the attributes. It is also possible that some attributes share the same value. Sample onfigurations are shown below.

$$A_1V_1, ..., A_nV_n | V_1A_1, ..., V_nA_n$$

$$V_1...V_kAV_{k+1}...V_n$$

$$A_1, A_2, ..., A_nV | V A_1, A_2, ..., A_n.$$

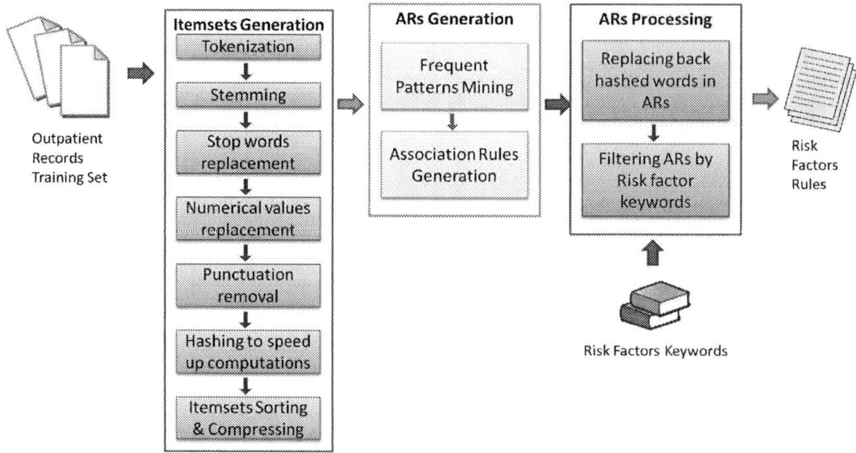

Figure 2: Risk Factors Association Rules Generation

Thus, when searching for frequent patterns, we consider a window of more than 10-12 words around each attribute. The rich terminology and flexible syntax structure hinder the application of traditional methods for extraction of collocations with gaps. Usual collocation extraction approaches would rather find the OR clishe phrases as collocations with highest frequency, moreover many A-V pairs would be erroneously considered as n-grams. Some FPs are given below.

> **E.g.: Positive examples**:
> общо състояние (*general condition*)
> щитовидна жлеза (*thyroid gland*)
> **Negative examples**:
> удължен експириум (*prolonged expiratory time*)
> има кашлица (*has a cough*)

Therefore we treat documents as bag of words rather than sequences, they are transformed to itemsets with single word occurrences only.

Given a set of pids S, support of an itemset I is the number of pids in S that contain I. We denote it as $supp(I)$. We define a threshold called *minsup* (minimum support). Frequent itemset (FI) I is one with at least minimum support count, i.e. $supp(I) \geq minsup$. The task of FPM of S is to find all possible frequent itemsets in S.

Most FPM algorithms generate all possible frequent patterns (FPs). The search space grows exponentially with the size of W. Summarised information for data relations can be extracted as maximal frequent itemsets (MFI). The condensed information not only accelerates the process, reducing redundancy, but also decreases significantly the number of frequent patterns for post-analysis.

An implication in the form $I \Rightarrow J$ is called *association rule*, where $I \subset W, J \subset W, I \cap J = \emptyset$. I is called antecedent and J is called consequent. Support of a rule is the number of pids in S that contain $I \cup J$, i.e.

$$sup(I \Rightarrow J) = sup(I \cup J) = P(I \cup J).$$

If $C\%$ of patient documents in S that contain I, contain also J, then the association rule $I \Rightarrow J$ holds with *confidence C* in S, i.e. this is the condition probability

$$conf(I \Rightarrow J) = P(J|I) = \frac{sup(I \cup J)}{sup(I)}.$$

The task of ARs mining in collection S is to generate all ARs with confidence above the user defined confidence (*minconf*) and support above user defined support (*minsup*). Rules that satisfy both a *minsup* and *minconf* are called strong. However, even for reasonable values of *minsup* and *minconf*, big datasets yield huge amounts of strong ARs. So we use an additional filter called *lift* that is defined as the ratio of the confidence of the rule and the confidence of its consequent.

$$lift(I \Rightarrow J) = \frac{P(I \cup J)}{P(I)P(J)}.$$

The lift represents the strenght of the relation between the consequent and its antecedent. Lift value < 1 indicates independence between them. Lift value > 1 means that the antecedent and consequent appear together more often than expected, i.e. are correlated. Such rules are potentially usefull for predicting the consequent in new sets.

For ARs generation we use algorithms for mining all association rules with the lift measure in a transaction database (Agrawal and Srikant, 1994) with implementation at SPMF[5]. For experiments is used algorithm for All Association Rule with FPGrowth with lift (Han et al., 2004). Let the two sets of generated ARs for *SHa* and *SHh* correspondingly be *ARa* and *ARh*.

3.2.3 Risk Factors Association Rules Filtering

In order to identify ARs for risk factors we use small lexicon with some keywords - $K = \{k_1, ..., k_m\}$. We convert back the hashed items from the ARs into words and obtain set *ARW*. For the two sets of ARs - *ARa* and *ARh* we have *ARaW* and *ARhW*. Thus the results ARs contain words. We filter those ARs that contain some of the keywords from the lexicon by projection.

$$ARaW_k = \{I \Rightarrow J | I \Rightarrow J \in ARaW \wedge \exists k \in K, k \in I \vee k \in J\}$$

$$ARhW_k = \{I \Rightarrow J | I \Rightarrow J \in ARhW \wedge \exists k \in K, k \in I \vee k \in J\}$$

3.3 Preprocessing of the test sets

Let *ST* be the test set of ORs. All Anamnesis (Patient History) sections formed the text collection *STa*, and all ORs Status texts - the collection *STh*. We process *STa* and *STh* separately. Similarly to the processing of the training set SH, we apply for *STa* and *STh* the first text analysis step - Itemsets Generation - but exclude the last procedures for hashing, compression and sorting.

3.4 Risk Factors Association Rules matching on the test sets

We match the corresponding type ARs to the test collections, i.e. ARs generated from the Anamnesis texts are mapped onto test collections that contain pids for Anamnesis, and the ARs generated from the Status parts of the ORs are mapped onto test collections that contain pids for Status. The result sets contain pids of patients at potential risk of chronic disease *H*.

$$RHa_k = \{p | p \in STa, I \Rightarrow J \in ARaW_k, I \subseteq p \wedge J \subseteq p\}$$

$$RHh_k = \{p | p \in STh, I \Rightarrow J \in ARhW_k, I \subseteq p \wedge J \subseteq p\}$$

[5]http://www.philippe-fournier-viger.com/spmf/index.php?link=algorithms.php

3.5 Structured information processing for patients at risk

Presence of some symptoms is a necessary but not sufficient condition for risk of chronic disease *H*. Some additional factors need further investigation, like related diagnosis with similar symptoms. We also need to study the other current diagnosis of the patient, to take into account age, gender, demographic information, etc. That's why we collect for each patient all pids from RHa_k and RHh_k and the associated structured information with the corresponding ORs from the test *ST*.

4 Experiments and Results

The chronic disease *H* that we investigate here is COPD (ICD-10 code J44), i.e. *H=J44*. The average prevalence of COPD in Bulgaria is 3.197% for 2014 among all Bulgarian citizens (Fig. 3). The average prevalence of both Schizophrenia (ICD-10 code F20) and Diabetes Melitus Type 2 (ICD-10 code E11) in Bulgaria is 0.688% for 2014 among all Bulgarian citizens (Fig. 4). However for 2014 the average prevalence of COPD among patients that suffer by both Schizophrenia and Diabetes Melitus Type 2 is relatively higher 5.576% than the average for the country (Fig. 5).

Some of the typical characteristics of COPD are: starting at middle age; symptoms develop slowly; prolonged smoking is a main reason; patients experience dyspnoea during physical efforts and significant irreversible airflow limitation. Thus in primary interest are ORs written by specialists: in Otolaryngology *(S14)*, Pulmology *(S19)* and Endocrinology *(S05)*. But we try to identify patients at risk, and probably some of them had no visits and consultations yet to such specialists. So we consider also collection of ORs for visits to general practitioners (GP) *(S00)*.

We have 4 text collections with ORs (Table 2): GP *(S00)*, Endocrinology*(S05)*, Otolaryngology *(S14)*, and Pulmology *(S19)*. We split these collections into training and test sets, depending on whether they are ORs for patients with *H=J44* or not. In addition we split them into two "*Anamnesis*" and "*Status*" sections of the ORs. Both sections are available for each patients so the training sets *SHa* and *SHh* contain the same number of pids. This is valid also for the test set *STa* and *STh* for each collection.

We can observe that for *SHh* (Table 3) the number of generated FPI and ARs is significantly

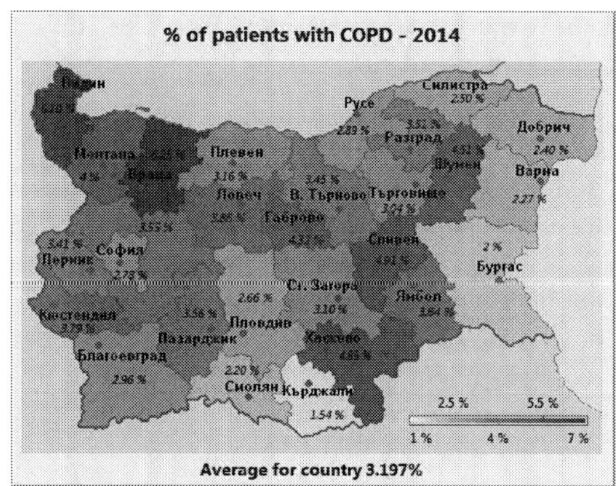

Figure 3: Prevalence of COPD (J44) in Bulgaria, 2014

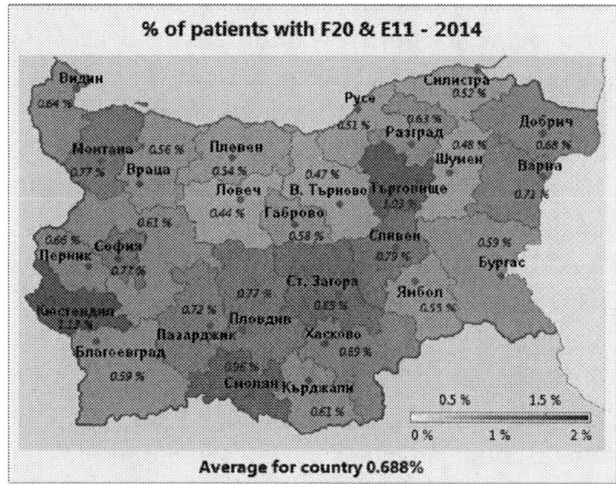

Figure 4: Prevalence of Schizophrenia (F20) and Diabetes Melitus Type 2 (E11) in Bulgaria, 2014

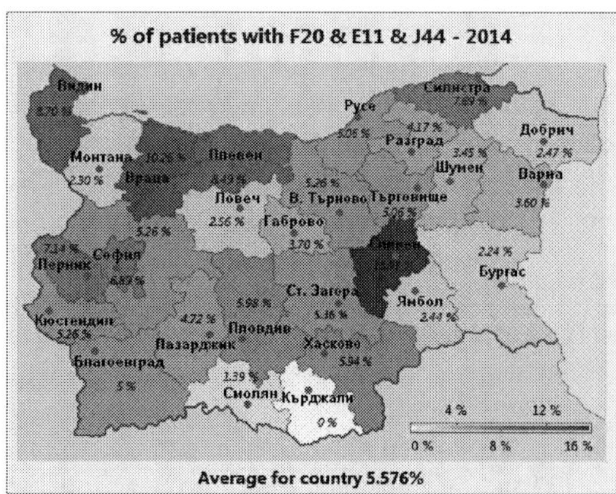

Figure 5: Prevalence of COPD (J44) among patients with both Schizophrenia (F20) and Diabetes Melitus Type 2 (E11) in Bulgaria, 2014

Year	2012	2013	2014	Total
patients	2,929	3,093	3,217	4,080
S00	45,402	46,238	51,894	143,534
S05	2,854	2,900	3,071	8,825
S14	368	351	396	1,115
S19	252	267	344	863

Table 1: Collection SD for patients with both Schizophrenia and Diabetes Melitus Type 2

Year	2012	2013	2014	Total
patients	144	166	179	293
S00	3,783	3,796	4,208	11,787
S05	253	273	262	788
S14	45	47	64	156
S19	158	172	202	532

Table 2: Training sets SHa and SHh

Set	ARa	FPI	minsup	$ARWa_K$
S00a	647	1,713	0.01	10
S05a	1,695,130	23,677	0.03	0
S14a	82,802	2,499	0.03	34
S19a	278,379	5,431	0.03	249,221

Table 3: Generated Association Rules for Anamnesis with minconf = 1.0 and minlift = 1.05

Set	ARh	FPI	minsup	$ARWh_K$
S00h	1,888,641	286,357	0.08	2
S05h	1,779,462	101,320	0.07	1,264
S14h	1,818	649	0.04	0
S19h	113,718	26,341	0.04	98,185

Table 4: Generated Association Rules for Status with minconf = 1.0 and minlift = 1.1

higher than for SHa (Table 4) even for higher *minsup* values, because the text in Status section is more coherent and contrain less variety of syntax structures. However the projection of these ARs to the keywords set K shrinks all the ARs sets in some cases to the ground. And it is not surprice that the majority of the filtered ARs comes from S19a and S19h - ORs from Pulmology.

The keywords for symphtoms of *J44* are:

K = {тежест, задух, кашлица, хрипове, храчки, умора, умораемост физическа, сърцебиене, трудно, експекторация, експириум} (*Weight, Breathlessness, Cough, Wheezing, Sputum, Fatigue, Tiredness, Physical, Palpitations, Difficult, Expectoration, Expiratory*).

Some generated ARs for COPD risk factors are:

умора експекторац => кашлиц SUP: 17 LIFT: 1.60
(Fatigue Expectoration => Cough)
храчки лесна умора => кашлиц SUP: 17 LIFT: 1.60
(Sputum Easy Fatigue => Cough)
храчки експекторац => задух SUP: 20 LIFT: 2.30
(Sputum Expectoration => Breathlessness)

Patients with potential risk of COPD are identified after matching the filtered rules of AWa_K and AWh_K to STa and STh correspondingly. The total number of ARs matches over the test sets of ORs is shown on (Table 5) and (Table 6) respectively.

$ARWa_k$	$ARW00a_k$	$ARW14a_k$	$ARW19a_k$
S00Ta	144	1,069	1,154
S05Ta	4	52	96
S14Ta	0	420	0
S19Ta	0	86	464
ORs	20	601	1,018

Table 5: COPD risk factors found in Anamnesis

$ARWa_k$	$ARW00h_k$	$ARW05h_k$	$ARW19h_k$
S00Th	0	3,086,665	829,995
S05Th	0	347,490	125,479
S14Th	0	0	0
S19Th	0	7,806	425,769
ORs	0	33,545	73,485

Table 6: COPD risk factors found in Status

In the following OR excerpt, items from the AR antecedent are highlighted in light blue color and the predicted consequent items are highlighted in pink color.

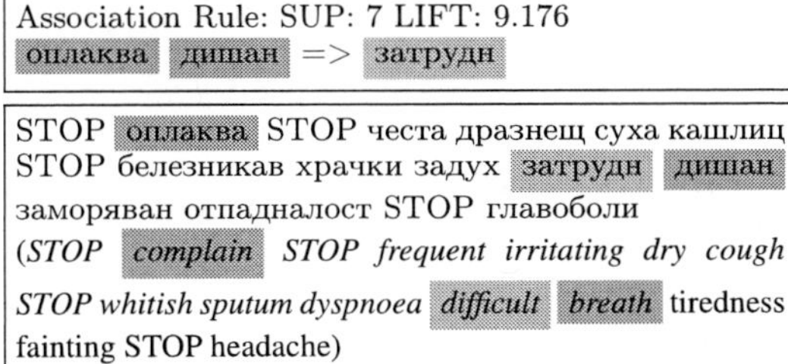

Patients that needs to be alerted for COPD risk factors are selected after analyses of some structured information in the ORs: age, gender, demographic region, etc.

COPD develops slowly and usually patient with age above 40s are at a higher risk . Risks are gender specific as well due to the prevalence of male (6.17%) vs. female (5.25%) patients. Demographic information helps to identify patient who live in regions with pollution, close to thermal power stations, etc. On Fig. 5 we can see that such regions in Bulgaria are around the town of Sliven (15.91%), Vidin (8.70%) and Vratsa (10.26%) in comparison with the average prevalence of COPD in the collection 5.576%. Another risk factor that needs further analysis is the patient smoking status because smoking is one of the major causes for COPD development. Some diagnoses related to the CORD symptoms are the following (with the corresponding ICD-10 codes in the parenthesis): *Asthma(J45), Status asthmaticus (J46), Congestive heart failure (I50.0), Bronchiectasis (J47), Tuberculosis (A15-A19), Bronchitis (J40-J42), Acute bronchiolitis (J20-J22), Emphysema (J43)*. So when planning alerts for patients at risk, one should check whether he/she has some of the diagnosis listed above and exclude those patient from the set *RH* for patients with risk alert.

5 Conclusion and Further Work

Here we show how to construct in a reliable manner a "could" of words signalling risks. This is important for a language like Bulgarian where no electornic linguistics resources of medical terminology are available. The existing very large archive of pseudonymised ORs, a nation-wide collection for 2010-2016, enables unique opportunities to acquire automatically lexical resources organised around names of diseases, medical conditions and/or specific groups of patients. The careful pre-selection of training corpora facilitates the explication of association rules; in this experiment we are aware about the comorbidity of CORD and Schizophrenia therefore we extract ORs for a cohort of patients which contains more CORD cases.

Despite the over-generation of ARs, the top rules are a reliable source of information which is easy to filter.

Another important achievement is the sketch of a clear procedure for discovery of patients at risk and issuing alerts to the healthcare authorities who need to take care about their implementation.

Future work involves processing of more complex linguistic constructions (negation) and considering typical risk factors (smoking).

Acknowledgments

This research is supported by the grant Special-IZed Data MIning MethoDs Based on Semantic Attributes (IZIDA), funded by the Bulgarian National Science Fund in 2017–2019, and the project DFNP-100/04.05.2016 "Automatic analysis of clinical text in Bulgarian for discovery of correlations in the Diabetic Registry" funded by the Bulgarian Academy of Sciences in 2016-2017. The team acknowledges the support of Medical University – Sofia, the Bulgarian Ministry of Health and the Bulgarian National Health Insurance Fund.

References

Rakesh Agrawal and Ramakrishnan Srikant. 1994. Fast algorithms for mining association rules in large databases. In *Proceedings of the 20th International Conference on Very Large Data Bases*. Morgan Kaufmann Publishers Inc., San Francisco, CA, USA, VLDB '94, pages 487–499. http://dl.acm.org/citation.cfm?id=645920.672836.

Nai-Wen Chang, Hong-Jie Dai, Jitendra Jonnagaddala, Chih-Wei Chen, Richard Tzong-Han Tsai, and Wen-Lian Hsu. 2015. A context-aware approach for progression tracking of medical concepts in electronic medical records. *Journal of biomedical informatics* 58:S150–S157.

Wendy W Chapman, Will Bridewell, Paul Hanbury, Gregory F Cooper, and Bruce G Buchanan. 2001. A simple algorithm for identifying negated findings and diseases in discharge summaries. *Journal of biomedical informatics* 34(5):301–310.

Joshua C Denny, Plomarz R Irani, Firas H Wehbe, Jeffrey D Smithers, and Anderson Spickard III. 2003. The knowledgemap project: development of a concept-based medical school curriculum database. In *AMIA Annual Symposium Proceedings*. American Medical Informatics Association, volume 2003, page 195.

Stefan Gindl. 2006. Negation detection in automated medical applications. *Vienna: Vienna University of Technology* .

Sergey Goryachev, Margarita Sordo, and Qing T Zeng. 2006. A suite of natural language processing tools developed for the i2b2 project. In *AMIA Annual Symposium Proceedings*. American Medical Informatics Association, volume 2006, page 931.

Jiawei Han, Jian Pei, Yiwen Yin, and Runying Mao. 2004. Mining frequent patterns without candidate generation: A frequent-pattern tree approach. *Data mining and knowledge discovery* 8(1):53–87.

Rave Harpaz, Alison Callahan, Suzanne Tamang, Yen Low, David Odgers, Sam Finlayson, Kenneth Jung, Paea LePendu, and Nigam H Shah. 2014. Text mining for adverse drug events: the promise, challenges, and state of the art. *Drug safety* 37(10):777–790.

Jitendra Jonnagaddala, Siaw-Teng Liaw, Pradeep Ray, Manish Kumar, Nai-Wen Chang, and Hong-Jie Dai. 2015. Coronary artery disease risk assessment from unstructured electronic health records using text mining. *Journal of biomedical informatics* 58:S203–S210.

Kory Kreimeyer, Matthew Foster, Abhishek Pandey, Nina Arya, Gwendolyn Halford, Sandra F Jones, Richard Forshee, Mark Walderhaug, and Taxiarchis Botsis. 2017. Natural language processing systems for capturing and standardizing unstructured clinical information: a systematic review. *Journal of Biomedical Informatics* .

Katherine P Liao, Tianxi Cai, Guergana K Savova, Shawn N Murphy, Elizabeth W Karlson, Ashwin N Ananthakrishnan, Vivian S Gainer, Stanley Y Shaw, Zongqi Xia, Peter Szolovits, et al. 2015. Development of phenotype algorithms using electronic medical records and incorporating natural language processing. *bmj* 350:h1885.

Yuan Luo, William K Thompson, Timothy M Herr, Zexian Zeng, Mark A Berendsen, Siddhartha R Jonnalagadda, Matthew B Carson, and Justin Starren. 2017. Natural language processing for ehr-based pharmacovigilance: A structured review. *Drug Safety* pages 1–15.

Amber Stubbs, Christopher Kotfila, Hua Xu, and Özlem Uzuner. 2015. Identifying risk factors for heart disease over time: Overview of 2014 i2b2/uthealth shared task track 2. *Journal of biomedical informatics* 58:S67–S77.

Dimitar Tcharaktchiev, Sabina Zacharieva, Galia Angelova, S. Boytcheva, Z. Angelov, P. Marinova, G. Nentchovska, L. Maneva, A. Velitchkov, G. Petrova, K. Koprivarova, I. Stoeva, M. Boyanov, R. Savova, R. Radev, L. Stoykova-Tchorbanova, E. Tasheva, P. Dentcheva, E. Foteva, K. Slavtcheva, B. Stoyanov, A. Stoev, S. Alexieva, E. Kotova, I. Kovatcheva, and T. Tomov. 2015. Building a bulgarian national registry of patients with diabetes mellitus. *Bulgarian Journal of Social Medicine* 2:19–21.

Özlem Uzuner, Xiaoran Zhang, and Tawanda Sibanda. 2009. Machine learning and rule-based approaches to assertion classification. *Journal of the American Medical Informatics Association* 16(1):109–115.

POMELO: Medline corpus with manually annotated food-drug interactions

Thierry Hamon[1,2], **Vincent Tabanou**[3], **Fleur Mougin**[4], **Natalia Grabar**[5], **Frantz Thiessard**[3,4]

[1] LIMSI, CNRS, Université Paris-Saclay, Orsay, France
[2] Université Paris 13, Sorbonne Paris Cité, Villetaneuse, France
[3] CHU de Bordeaux, Pole de sante publique, Service d'information médicale, Bordeaux, France
[4] Univ. Bordeaux, Inserm, Bordeaux Population Health Research Center,
team ERIAS, UMR 1219, F-33000 Bordeaux, France
[5] CNRS UMR 8163 STL, Université Lille 3, 59653 Villeneuve d'Ascq, France

`hamon@limsi.fr, vincenttabanou@aol.com, fleur.mougin@u-bordeaux.fr,`
`natalia.grabar@univ-lille3.fr, frantz.thiessard@u-bordeaux.fr`

Abstract

When patients take more than one medication, they may be at risk of drug interactions, which means that a given drug can cause unexpected effects when taken in combination with other drugs. Similar effects may occur when drugs are taken together with some food or beverages. For instance, grapefruit has interactions with several drugs, because its active ingredients inhibit enzymes involved in the drugs metabolism and can then cause an excessive dosage of these drugs. Yet, information on food/drug interactions is poorly researched. The current research is mainly provided by the medical domain and a very tentative work is provided by computer sciences and NLP domains. One factor that motivates the research is related to the availability of the annotated corpora and the reference data. The purpose of our work is to describe the rationale and approach for creation and annotation of scientific corpus with information on food/drug interactions. This corpus contains 639 MEDLINE citations (titles and abstracts), corresponding to 5,752 sentences. It is manually annotated by two experts. The corpus is named *POMELO*. This annotated corpus will be made available for the research purposes.

1   Introduction

Prescribed medicines depend on initial marketing authorization to guarantee the security of patients. Nevertheless, medicines can cause adverse drug reactions (ADRs) discovered during clinical trials, but usually later, in a pharmacovigilance context, while drugs are administered to patients (Aagaard and Hansen, 2013; Brahma et al., 2013; Cote and Choy, 2013; Yom-Tov and Gabrilovich, 2013; Wei et al., 2013). For this reason, prescription and intake of drugs is controlled all over their marketing and use by patients.

When patients take more than one medication, they may be at risk of drug interactions, which means that a given drug can cause unexpected effects when taken in combination with other drugs. For instance, sedative and pain medication cause an important drowsiness in patients, while other drugs (*benzylpenicillin* and *heparin*) interact between them and cannot be placed in the same syringe. Most drugs are not concerned by such effects. Yet, when such adverse events happen, they may have negative and serious effects on patients and their health. Hence, it is important to know the possible interactions between drugs and to clearly indicate them to patients and to medical staff. For this reason, for several years now, automatic extraction of drug-drug interactions is heavily researched in order to provide an updated and timely information on known interactions between drugs or between their active principles. In the same way, it becomes possible to discover potential adverse effects and adverse reactions (Aronson and Ferner, 2005) of these drugs.

A different but yet related situation occurs when drugs are taken together with certain food or beverages. Food and drug interaction may also lead to negative effects on health and well-being of patients. For instance, grapefruit has interactions with several drugs, because its active ingredients inhibit enzymes involved in the drugs metabolism and can then cause an excessive dosage of these drugs (Duke Med Health News, 2013; Greenblatt and Derendorf, 2013). Due to the difficulty to detect them, because patients usually do not remem-

Proceedings of the Biomedical NLP Workshop associated with RANLP 2017, pages 73–80,
Varna, Bulgaria, 8 September 2017.

ber the food they have taken, and to their complexity, such situations are studied less frequently, even if they are important for patients and for the medical care process. Our main interest is to study food-drug interactions and to automatically detect them in scientific literature.

Most information on food/drug interactions is recorded in unstructured sources, such as scientific articles and some knowledge bases, like Drug-Bank[1] (Wishart et al., 2006), or possibly in discussion fora which provide patient point of view of adverse events. Yet, this information remains poor. For instance, DrugBank records textual information about food/drug interactions for less than 10% of drugs, and it mainly provides information on optimal drug intake time.

Regarding these observations, our objective is to use and mine scientific bibliographical data in order to describe interactions that exist between drugs and food, and that may lead to adverse effects. To achieve this goal, our first step is to design and to annotate a dataset of MEDLINE abstracts. This is the purpose of the work presented in this paper.

In what follows, we first present some related work (section 2). We then present the annotation scheme (section 3), and describe the corpus definition and the annotation process (sections 4 and 5). Finally, we present our results (section 6), and discuss our work and conclude with future orientations (section 7).

2 Related work

We present two kinds of works: those performed by pharmacists and pharmacovigilance experts on drug/drug interactions (DDI) and food/drug interactions (FDI) in medical domain, and those performed by computer scientists on information extraction.

2.1 Medical domain

In has been defined that interactions can occur in different ways. The interactions presented here have been defined for the DDI cases, but they show very similar effects when the FDIs occur. Hence, two drugs given together may act at the same or similar receptor, which can lead to a greater or to a decreased effect of either drug. Another situation is when one drug is affected by action of another drug. In this case, their absorption, distribution, metabolism or excretion (commonly called ADME) are involved (Doogue and Polasek, 2013). We give here some examples of DDIs found online[2]:

- *Absorption.* Some drugs can alter the absorption of another drug. For example, *calcium* can block absorption of some medications. Hence, the HIV treatment *dolutegravir (Tivicay)* should not be taken at the same time as *e.g. calcium carbonate (Tums, Maalox)*, because it can lower the amount of *dolutegravir* absorbed and reduce its effectiveness in treating HIV infection. For the same reason, many drugs cannot be taken with milk or dairy products because they will bind with the calcium;

- *Distribution.* Protein-binding interactions can occur when highly protein-bound drugs compete for a limited number of binding sites. One example is between *fenofibric acid (Trilipix)*, used to lower cholesterol in the blood, and *warfarin*, a common blood thinner to help prevent clots. *Fenofibric acid* can increase the effects of *warfarin* and cause the bleeding in patient;

- *Metabolism.* Drugs are usually eliminated from the body further to their changes through metabolism. Enzymes in the liver, usually the CYP450 enzymes, are often responsible for breaking down drugs and for their elimination from the body. However, enzyme levels may go up or down and affect how drugs are broken down. For example, using *diltiazem*, a blood pressure medication, with *simvastatin*, a medicine to lower cholesterol, may elevate the blood levels and cause side effects due to *simvastatin*. Indeed, *diltiazem* can block the CYP450 3A4 enzymes needed for the breakdown of *simvastatin*, in which case, high blood levels of *simvastatin* can lead to serious liver and muscle side effects. Another example is when grapefruit affects the action of the CYP3A4 enzyme, thus also affecting the intake of several drugs (Duke Med Health News, 2013; Greenblatt and Derendorf, 2013);

---

[1] http://www.drugbank.ca/

[2] https://www.drugs.com/drug_interactions.html

74

- *Excretion*. Some *nonsteroidal antiinflammatory drugs (NSAIDs)* (e.g. *indomethacin*), may lower kidney function and affect the excretion of *lithium*, a drug used for bipolar disorder, in which case its action can be increased.

From these examples, we can highlight several points related to the intake of drugs:

- when several medications are taken together, patients should define the best way to take them (*e.g.* time, dose) with their doctor;

- DDIs and FDIs may show similar action patterns because they may contains same or similar active principles, like shown above with the *calcium* intake, present in drugs and in diary products, or with grapefruit and *diltiazem* changing the behavior of some enzymes;

- the interaction can follow several patterns: action of a given medication can be decreased, increased or cause side effects which are usually not observed.

2.2 Computer Sciences

As noticed above, it is important to research the issues related to DDIs and FDIs. Concerning the DDIs, and globally the ADRs (Adverse Drug Reactions), their reporting is extremely low. For instance, in France, 96% of the ADRs are simply not reported (Moride et al., 1997; Lacoste-Roussillon et al., 2001). As for the FDIs, they are even more difficult to identify for several reasons: the large number of possible interactions, the difficulty of describing meals in a standard ADR reporting form, the difficulty to remember exactly which food has been taken at a given moment, and probably also for sociological reasons because drugs and pathology are connected unconsciously and associated with negative feelings, whereas food is rather an indication of good health. For these reasons, it is important to provide automatic NLP (Natural Language Processing) methods for mining available sources of information, like MEDLINE bibliographical database[3].

Several works have been done on automatic extraction of DDIs (Duda et al., 2005; Björne et al., 2013; Ayvaz et al., 2015; Kim et al., 2015; Kolchinsky et al., 2015; Liu et al., 2016; Schneider and Boyce, 2016). In most cases, supervised categorization methods are exploited for the detection of entities and of their interactions. This explains why the majority of these works are part of the NLP challenges, like *SEM (Segura-Bedmar et al., 2013) for DDI extraction and BIOCREATIVE (Krallinger et al., 2009) for PPI (Protein–Protein Interaction) extraction. Indeed, the *SEM challenge proposes task dedicated to DDI extraction and provides annotated corpora. The main contribution of our work is related to the creation of biomedical corpora and their annotation with information on food/drug interactions.

Notice that there is very little work on automatic FDI extraction. Currently, several knowledge bases, semantic resources and repositories concerning the involved entities (i.e., drugs, food and diseases) are available (Brown et al., 1999; NLM; RxNorm; Kuhn et al., 2010). Yet, the FDI information is fragmented and scattered across these bases and repositories. As consequence, there is no explicit relations between these entities. Linked Data projects, such as Linked Open Drug Data (LODD[4]), attempt to create fine-grained links between such knowledge bases. In addition, the already mentioned DrugBank (Wishart et al., 2006) knowledge base contains various kinds of information on drugs, although their relations with food is provided as free-text fields and is mainly concerned with the optimal drug intake time. The effort done for the formalization of the FDI information from DrugBank (Jovanovik et al., 2014) has been performed manually.

In the next sections, we propose the description of the methodology for the creation of annotated corpus with FDIs.

3 Annotation schema: Representation of drug related information and of FDIs

In Figure 1, we present the model of the drug-related information. The model is instantiated for the *solumedrol* medication. Hence, a given drug has an international name (*DCI*) and a therapeutic class (or is-a relation). It has a composition and is prescribed for specific indications and with specific features (*e.g.* dosage, mode, frequency and duration of administration). Then, a drug may have adverse effects, including those due to action of food (FDIs) or of other drugs (DDIs).

The annotated entities must belong to one of these categories:

---

[3]https://www.ncbi.nlm.nih.gov/pubmed

[4]http://www.w3.org/wiki/HCLSIG/LODD

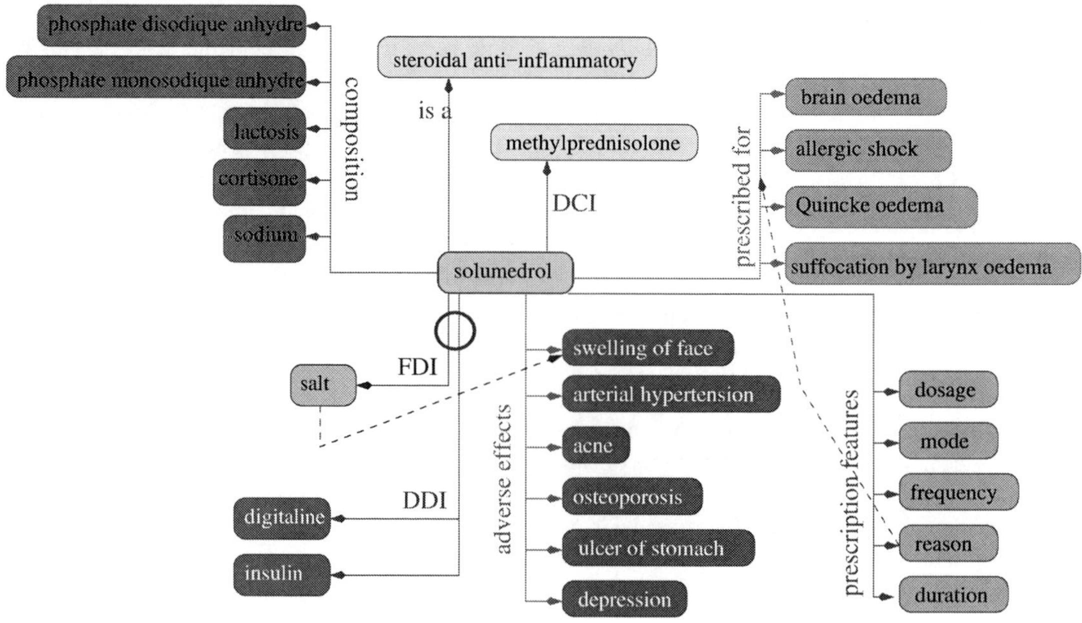

Figure 1: Model of the drug-related information

1. food names and their different types (ingredients, cooked meals, food supplements...),

2. meal time (before, during, after),

3. drug names and related information (dosage, frequency, duration, mode),

4. disorders for which a given drug is indicated,

5. side effects (including FDIs) of drugs.

According to types of actions of drugs on patients and to the ADME model (section 2.1), we propose to investigate the following types of interactions between these entities:

- decrease, reduce, slow down or make disappear drug effect (absorption, elimination...) due to food,

- increase or speed up drug effect due to food,

- make appear, have (new) side effects, make appear negative effect, or worsen drug effect due to food,

- increase side effect of drugs due to food,

- improve drug effect, have positive effect on drug, or reduce side effect of drug,

- treat a disorder,

- have no effect on drug,

- drug must be taken without food,

- have effect on drug or general relation with drug. These relations are under-defined: relation or effect exist but it is not possible to decide what kind of relation or effect it is.

4   Corpus design

The MEDLINE bibliographical base has been queried in order to extract citations related to food-drug interactions with the following query:

("FOOD DRUG INTERACTIONS"[MH] OR "FOOD DRUG INTERACTIONS*") AND ("adverse effects*")

In December 2013, it permitted to obtain a set of 639 citations, of which we exploit titles and abstracts. This corpus is called *POMELO*, namely *grapefruit* in French, because has been built and annotated during the French MESHS-funded project POMELO.

5   Corpus annotation

The POMELO corpus (639 titles and abstracts corresponding to 5,752 sentences) has been manually annotated in order to make explicit the information on food/drug interactions. Two experts have been involved in the annotation process: one resident and one medical doctor. The annotation has been done mainly by the resident, who was helped by the medical doctor when facing difficult situations. The annotation has been performed with the BRAT software (Stenetorp et al., 2012).

To prepare the annotation, we exploited some existing resources in English and French, which have been automatically projected on corpus. For instance, the food has been pre-annotated using:

- the USDA National Nutrient Database[5]

- the Codex Alimentarius of the WHO (World Health, Organization)[6],

- and resources built from some recipes.

Other entities have been pre-annotated with existing terminologies: disorders and side effects (Brown et al., 1999; NLM; Kuhn et al., 2010), and drugs (RxNorm; Wishart et al., 2006). Besides, specific resources have been built for the annotation of dosage, frequency, duration and mode of drug administration.

Then, the POMELO corpus has been checked out for the correctness of entities and further annotated with relations by the annotators.

6 Results

In Figure 2, we give an example of an annotated citation. Drugs are in blue, food in green and adverse effects in cyan. Other entities are related to dosage, frequency and duration. Then, relations between these entities are marked up.

---

[5] http://ndb.nal.usda.gov/ndb/search/list

[6] http://www.codexalimentarius.org

Entities	Nb
drug	4,953
food	2,783
treated disease	645
drug effect	558
side effect	1,985
meal time	1,027
mode	539
dosage	767
duration	86
frequency	282

Table 1: Types and number of the annotated entities

Relation	Nb
decrease absorption	64
slow absorption	21
slow elimination	18
increase absorption	52
speed up absorption	4
new side effect	4
negative effect on drug	91
worsen drug effect	16
has side effect	434
increase side effect	239
positive effect on drug	23
reduce side effect	23
improve drug effect	14
treat	350
no effect on drug	145
without food	23
has effect	233
relation	706

Table 2: Types and number of the annotated relations

In Tables 1 and 2, we indicate the types and numbers of entities and relations manually annotated by experts. We can see for instance that among the most frequent relations we can find:

- drugs `have` side effects (n=434):
  Atovaquone suspension was well tolerated; diarrhea, nausea, fatigue, and rash were the most common adverse events.

- drugs `treat` disorders (n=350):
  Metrifonate is an inhibitor of cholinesterase effective in the treatment of Alzheimer's disease.

- food `increases` side effects (n=239):
  Animals infused ethanol-containing diets adequate in carbohydrate developed steatosis, but had no other signs of hepatic pathology.

- food `has no effect` on drugs (n=145):
  Azimilide dihydrochloride may be orally administered to patients without regard to the prandial state.

- food `has undefined effect` on drugs (n=233):
  In both studies the equivalence in AUC of DDVP was paralleled by equivalent effects on BChE inhibition.

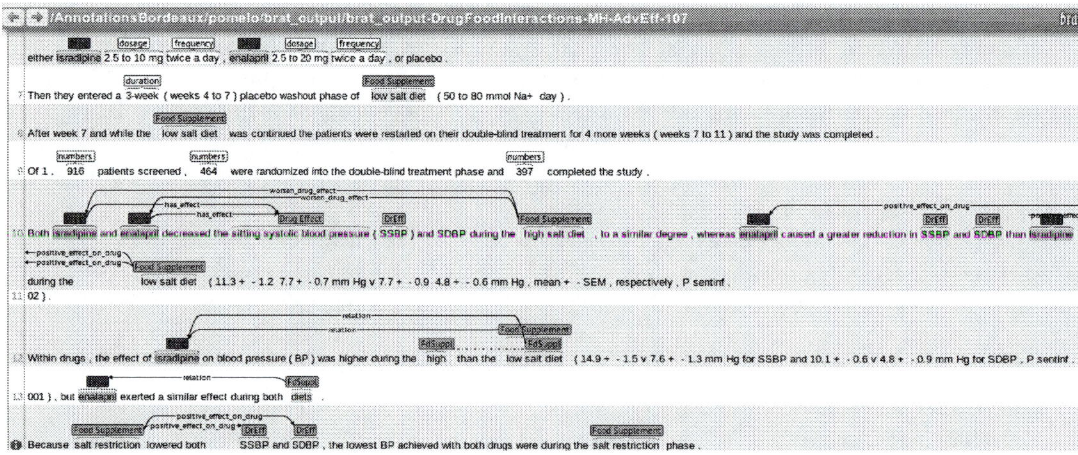

Figure 2: Sample of annotated document (title and abstract) issued from a Medline citation

- food `has undefined relation` with drugs (n=706):
  We know that changing a customary diet to one high in protein and low in carbohydrate increases the rates of metabolism of antipyrine and theophylline, and shifting to an isocaloric diet of low protein- protein-high carbohydrate slows the rates of metabolism of these drugs.

This corpus will be made available for the research purposes. In this way, we expect to encourage the research on food/drug interactions.

7 Conclusion

We described our work done in order to create a corpus of biomedical literature (titles and abstracts) annotated with information on food/drug interactions. The corpus is called *POMELO*, namely *grapefruit* in French. Titles and abstracts are obtained from MEDLINE bibliographical base. We first propose a model of the food/drug-related information, which takes into account various aspects going from composition of drugs to their intake and possible adverse effects. Then, the annotation is performed by two experts: one resident and one medical doctor. Several entities are annotated, such as drugs, food, diseases and drug effects, meal time, and drug-related information (mode, dosage, duration and frequency). These entities are pre-annotated automatically and then checked out manually by the annotators. Then, relations between these entities are manually annotated. Among the most frequent relations, we can observe for instance: drugs `have` side effects, drugs `treat` disorders, food `increases` side effects, food `has no effect` on drugs, food `has undefined effect` on drugs, food `has undefined relation` with drugs.

We have several orientations for the future work on this research. One of the orientations is concerned by increasing the size and quality of the annotated corpus: (1) an updated MEDLINE query indicates that currently there are more citations indexed with the queried keywords; (2) another related MESH keyword (*herb/drug interactions*) can be exploited to enrich the corpus; (3) even if two experts have been involved in the annotation, the annotation can be done by other independent annotators; (4) finally, this English-language annotated corpus can be enriched with French-language citations and documents. Another orientation is related to the exploitation of this annotated corpus: (1) use of the annotations for creating the model for automatic extraction of food/drug interactions; (2) exploit this model for a systematic extraction of FDIs and their recording together with their evidence level; (3) creation of a knowledge base with food/drug interactions and their use by medical professionals and patients.

The *POMELO* corpus will be made available for the research purposes on the web site of the MIAM project (`https://miam.limsi.fr/`).

Acknowledgments

This work was supported by the MESH emergent project and Agence Nationale de la Recherche through the grant ANR-16-CE23-0012 France.

References

L Aagaard and EH Hansen. 2013. Adverse drug reactions reported by consumers for nervous system medications in europe 2007 to 2011. *BMC Pharmacol Toxicol* 14:30.

JK Aronson and RE Ferner. 2005. Clarification of terminology in drug safety. *Drug Saf* 28(10):851–70.

S Ayvaz, J Horn, O Hassanzadeh, Q Zhu, J Stan, NP Tatonetti, S Vilar, M Brochhausen, M Samwald, M Rastegar-Mojarad, M Dumontier, and RD Boyce. 2015. Toward a complete dataset of drug-drug interaction information from publicly available sources. *J Biomed Inform* 55:206–17.

J Björne, S Kaewphan, and T Salakoski. 2013. UTurku: Drug named entity detection and drug-drug interaction extraction using svm classification and domain knowledge. In *International Workshop on Semantic Evaluation (SemEval 2013)*. pages 1–9.

DK Brahma, JB Wahlang, MD Marak, and MCh Sangma. 2013. Adverse drug reactions in the elderly. *J Pharmacol Pharmacother* 4(2):91–4.

EG Brown, L Wood, and S Wood. 1999. The medical dictionary for regulatory activities (MedDRA). *Drug Saf.* 20(2):109–117.

GM Cote and E Choy. 2013. Role of epigenetic modulation for the treatment of sarcoma. *Curr Treat Options Oncol* .

MP Doogue and TM Polasek. 2013. The abcd of clinical pharmacokinetics. *Ther Adv Drug Saf* 4(1):5–7.

Stephany Duda, Constantin Aliferis, Randolph Miller, Alexander Slatnikov, and Kevin Johnson. 2005. Extracting drug-drug interaction articles from Medline to improve the content of drug databases. In *AMIA Symp*. pages 216–20.

Duke Med Health News. 2013. Grapefruit: enemy of many medications. in some patients, the interaction of fruit and drug may put their life and health at risk. *Duke Med Health News* 19(2):1–2.

DJ Greenblatt and H Derendorf. 2013. Grapefruit-medication interactions. *CMAJ* 185(6):507.

Milos Jovanovik, Aleksandra Bogojeska, Dimitar Trajanov, and Ljupco Kocarev. 2014. Inferring cuisine - drug interactions using the linked data approach. *Nature* 5(9346):1–7.

Sun Kim, Haibin Liu, Lana Yeganova, and W. John Wilbur. 2015. Extracting drug–drug interactions from literature using a rich feature-based linear kernel approach. *Journal of Biomedical Informatics* 55:23–30.

Artemy Kolchinsky, Anália Lourenço, Heng-Yi Wu, Lang Li, and Luis M. Rocha. 2015. Extraction of pharmacokinetic evidence of drug-drug interactions from the literature. *PLoS ONE* 10(5):1–24.

M Krallinger, F Leitner, and A Valencia. 2009. The biocreative ii.5 challenge overview. In *BioCreative II 5 Workshop 2009 on Digital Annotations*.

Michael Kuhn, Monica Campillos, Ivica Letunic, Lars Juhl Jensen, and Peer Bork. 2010. A side effect resource to capture phenotypic effects of drugs. *Molecular Systems Biology* 6(1).

C. Lacoste-Roussillon, P. Pouyanne, F. Haramburu, G. Miremont, and B. Bégaud. 2001. Incidence of serious adverse drug reactions in general practice: a prospective study. *Clin Pharmacol Ther* 69(6):458–462.

Shengyu Liu, Buzhou Tang, Qingcai Chen, and Xiaolong Wang. 2016. Drug-drug interaction extraction via convolutional neural networks. *Comput Math Methods Med. 2016; 2016: 6918381* 2016:1–8.

Y. Moride, F. Haramburu, A. Requejo Alvarez, and B. Bégaud. 1997. Under-reporting of adverse drug reactions in general practice. *Br J Clin Pharmacol* 43(2):177–181.

NLM. 2001. *Medical Subject Headings*. National Library of Medicine, Bethesda, Maryland. *www.nlm.nih.gov/mesh/meshhome.html*.

RxNorm. 2009. RxNorm, a standardized nomenclature for clinical drugs. Technical report, National Library of Medicine, Bethesda, Maryland. Available at `www.nlm.nih.gov/research/umls/rxnorm/docs/index.html`.

Jodi Schneider and Richard D. Boyce. 2016. Acquiring and representing drug-drug interaction knowledge as claims and evidence. In *NLM Informatics Training Conference*.

Isabel Segura-Bedmar, Paloma Martinez, and Maria Herrero-Zazo. 2013. SemEval-2013 task 9 : Extraction of drug-drug interactions from biomedical texts (DDIExtraction 2013). In *Lexical and Computational Semantics (*SEM)*. pages 341–350.

Pontus Stenetorp, Sampo Pyysalo, Goran Topić, Tomoko Ohta, Sophia Ananiadou, and Jun'ichi Tsujii. 2012. brat: a web-based tool for nlp-assisted text annotation. In *Proceedings of the Demonstrations at the 13th Conference of the European Chapter of the Association for Computational Linguistics*. Association for Computational Linguistics, Avignon, France, pages 102–107. http://www.aclweb.org/anthology/E12-2021.

Z Wei, C Doria, and Y Liu. 2013. Targeted therapies in the treatment of advanced hepatocellular carcinoma. *Clin Med Oncol* 7:87–102.

David S. Wishart, Craig Knox, An Chi Guo, Savita Shrivastava, Murtaza Hassanali, Paul Stothard, Zhan Chang, and Jennifer Woolsey. 2006. Drugbank: a comprehensive resource for in silico drug discovery and exploration. *Nucleic Acids Research* 34:668–672. Database issue.

E Yom-Tov and E Gabrilovich. 2013. Postmarket drug
surveillance without trial costs: discovery of adverse
drug reactions through large-scale analysis of web
search queries. *J Med Internet Res* 15(6):124–125.

Annotation of Clinical Narratives in Bulgarian language

Ivaylo Radev   Kiril Simov   Galia Angelova   Svetla Boytcheva
Institute of Information and Communication Technologies,
Bulgarian Academy of Sciences
radev@bultreebank.org, kivs@bultreebank.org,
galia@lml.bas.bg, svetla.boytcheva@gmail.com

Abstract

In this paper we describe annotation process of clinical texts with morphosyntactic and semantic information. The corpus contains 1,300 discharge letters in Bulgarian language for patients with Endocrinology and Metabolic disorders. The annotated corpus will be used as a Gold standard for information extraction evaluation of test corpus of 6,200 discharge letters. The annotation is performed within Clark system — an XML Based System for Corpora Development. It provides mechanism for semi-automatic annotation. First a pipeline for Bulgarian morphosyntactic annotation and a cascaded regular grammar for semantic annotation are run, then rules for cleaning of frequent errors are applied. At the end the obtained result is manually checked. Our goal is to adapt the morphosyntactic tagger to the domain of clinical narratives as well.

1   Introduction

Today the electronic patient records and clinical notes are a fast growing research resource of medical data. These free text documents written by physicians contain a lot of valuable medical information despite the fact that sensitive data makes them hard to work with.

In countries like Sweden, UK and US researchers have started to use the electronic health records (EHR) to create corpora for two main purposes – in order to perform information extraction for medical research and for training domain specific systems to cope with these texts. Related subtasks are: automated de-identification for research work with sensitive data; extraction of medical time-lines in case development, with identification of deceasse and treatment; doing information retrieval and text mining; performing research in order to find relationships between diagnoses, treatments etc.; creation of golden standard corpora for evaluation and training; name entity recognition and annotation.

In this paper we describe annotation with morphosyntactic and semantic information of clinical texts. The corpus contains 1,300 discharge letters in Bulgarian language for patients with Endocrinology and Metabolic disorders. The annotated corpus will be used for information extraction evaluation.

The paper is structured as follows. Section 2 overviews related work with focus on the technological solutions. Section 3 presents the method we use, section 4 – the experiments and results. Section 5 contains the conclusion and discusses future work.

2   Related Work

Relevant references discuss annotation projects for corpora of medical texts in various natural languages. Studying the literature we adapted some principles for our annotation, although the sources are not directly connected to Bulgarian language.

A variety of approaches are described in the literature: e.g. for temporal annotations; pipeline with lexical features to extract time and event mentions; statistical chunking system for annotation; pipeline of tools for automatic processing of clinical texts and tokenization through part-of-speech tagging and dependency parsing; a simplification system, for automated change and adjusting of the text in health records in order to make them easier to understand; biomedical entity recognition dataset using a human-into-the-loop approach. Here we enumerate some annotation approaches correspondingly to language layers.

Proceedings of the Biomedical NLP Workshop associated with RANLP 2017, pages 81–87,
Varna, Bulgaria, 8 September 2017.

Entities. The article (Ogren et al., 2007) reports about the construction of a gold-standard dataset consisting of annotated clinical notes suitable for evaluating a biomedical named entity recognition system. The dataset is the result of consensus between four human annotators and contains 1,556 annotations on 160 clinical notes using 658 unique concept codes from SNOMED-CT corresponding to human disorders. Inter-annotator agreement was calculated on annotations from 100 of the documents for span (90.9%), concept code (81.7%), context (84.8%), and status (86.0%) agreement. Another corpus is designed to support automatic recognition of symptoms in unseen text. It consists of clinical free text records enriched with annotation for symptoms of a particular disease (ovarian cancer). The data (approximately 192K words) was annotated by three clinicians and a procedure was devised to resolve disagreements. The corpus is allows also to investigate the amount of symptom-related information in clinical records that is not coded (Koeling et al., 2011). Recognising entities is related to de-identification of sensitive information; the definitions of annotation classes are not self-evident. The article (Dalianis and Velupillai, 2010) presents two refined variants of an annotated gold standard corpus for de-identification of patient records in Swedish, one created automatically, and one created through discussions among the annotators. These are used for the training and evaluation of an automatic de-identification system based on the Conditional Random Fields algorithm. Promising results are acheived for both Gold Standards: F-score around 0.80 for a number of experiments on 4-6,000 instances, with higher results for certain annotation classes. The construction of three annotated corpora is presented in (Deleger et al., 2012) that serve as gold standards for medical NLP tasks. The annotated narratives are clinical notes from the medical record, clinical trial announcements, and FDA drug labels. High inter-annotator agreements is reported; the corpora are made public to facilitate translational NLP tasks that require cross-corpora interoperability. An annotated corpus (PhenoCHF), focussing on the identification of phenotype information for a specific clinical sub-domain, i.e., congestive heart failure (CHF), is presented in (Alnazzawi et al., 2014). The corpus integrats information from both EHRs (300 discharge summaries) and literature articles (5 full-text papers). The annotation scheme, whose design was guided by a domain expert, includes both entities and relations pertinent to CHF. Two further domain experts performed the annotation with agreement rates up to 0.92 F-Score.

Syntax. The paper (Fan et al., 2013) presents the development of a corpus with syntactic annotation (treebank) with intention to handle ill-formed sentences which are common in clinical text. A supplement to the Penn Treebank II guidelines was developed for annotating clinical sentences. After three iterations of annotation and adjudication on 450 sentences, the annotators reached an F-measure agreement rate of 0.930 (while intra-annotator rate was 0.948) on a final independent set. A total of 1100 sentences from progress notes were annotated that demonstrated domain-specific linguistic features. A statistical parser retrained with combined general English (mainly news text) annotations and our annotations achieved an accuracy of 0.811 (higher than models trained purely with either general or clinical sentences alone). In (Savkov et al., 2016), an approach to training domain specialists with no linguistic background to annotate clinical text is presented. The authors describe a de-identified corpus of free text notes, a shallow syntactic and named entity annotation scheme. A statistical chunking system for such clinical text with a stable learning rate and good accuracy is presented, indicating that the manual annotation is consistent and that the annotation scheme is tractable for machine learning.

Semantics. The Clinical E-Science Framework (CLEF) project aims at the identification of semantic entities and relationships in clinical narratives. The CLEF corpus consists of clinical narratives, histopathology reports and imaging reports from 20 thousand patients. A subset of this corpus was selected for manual annotation of clinical entities and relationships (Roberts et al., 2007). By entity, some real-world thing referred to in the text is meant: the drugs that are mentioned, the tests that were carried out etc. The relationships between entities correspond to the condition indicated by a drug, the result of an investigation etc. Annotation is anchored in the text. Annotators mark spans of text with a type: drug, locus and so on. Annotators may also mark words that modify spans (such as negation), and mark relationships as links between spans. Two or more spans may refer to the same thing in the real world, in which case they co-refer. Each text was annotated by 2 experts independently. In total, 27 annotators are involved in debugging, annotation and review roles. They are drawn from practicing clinicians, medical informaticians, and final year medical students. This corpus was used as a gold standard prividing temporal links (called CTlinks) between TLCs (Temporally Located CLEF entities, which comprise investigations, interventions and conditions) and temporal expressions: dates and times (both absolute and relative), as well as durations, as specified in the TimeML TIMEX3 standard (Roberts et al., 2008). The gold standard is a resource against which to assess the Information Extraction (IE) results of CLEF system. In addition, statistical models of the text may be built by machine learning algorithms. In 2008 the authors write that "the annotated CLEF corpus is the richest resource of semantically marked up clinical text yet created". The semantic annotation scheme, the annotation methodology, and the distribution of annotations in the final corpus are detailed in (Roberts et al., 2009).

Discource and Standardization. The Ontology Development and Information Extraction corpus (ODIE) annotated anaphoric relations in clinical narratives. The gold standard annotations resulted in 7214 markables, 5992 pairs and 1304 chains. These early shared annotation resources revealed the lack of common annotation schemes and community adopted standards and conventions for normalization (Savova, 2017). Recent ambitious projects aim at the annotation of timelines, in order to enable natural language understanding by discovering events and their relations on a timeline. Temporal relations are of prime importance in biomedicine as they are intrinsically linked to diseases, signs and symptoms, and treatments. The annotation guidelines of THYME project ("Temporal Histories of Your Medical Events") are based on TIMEX3[1].

3 Methodology

The annotation is performed in two steps:

1. Automatic preprocessing and

2. Manual errors checking and correction.

The first step is done by BulTreeBank pipeline for Bulgarian (Savkov et al., 2012) updated with

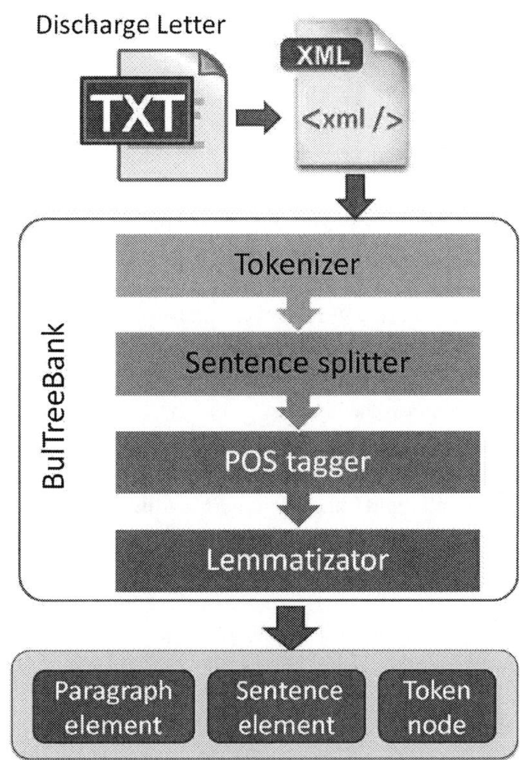

Figure 1: Automatic preprocessing

new tools — we substituted previous POS tagger and dependency parser with new ones based on MATE tool[2]. The process starts with simple discharge letter in text format written by the physician (Fig. 1). The text document is converted to XML format. After that we use tokenizer, sentence splitter, POS tagger and lemmatizator to automatically process the raw texts. The result from this processing includes the following information:

- ***Paragraph element (p)*** — contains some meta data like age, gender and location of the patient and the main sections of the discharge letter – anamnesis, health status, diagnosis, treatment, clinical exams, consultations, etc.

- ***Sentence element (s)*** — does not have additional information. Very hard to be done because the physicians neglect the punctuation rules.

- ***Token node (tok)*** — the main node of the tree. It has all the linguistic information like

---

[1] http://clear.colorado.edu/compsem/documents/THYME_guidelines.pdf

[2] http://www.ims.uni-stuttgart.de/forschung/ressourcen/werkzeuge/matetools.en.html

POS and lemmas. Also it has the term attribute.

The overall performance accuracy of the original pipeline droped significantly due to the reach medical terminology included in the texts. The result XML documents are after that checked and annotated further manually. During this process we are using CLaRK system[3] — an XML Based System for Corpora Development (Simov et al., 2001), (Simov et al., 2004). The core of CLaRK is an Unicode XML Editor, which is the main interface to the system. Via it the user could edit, search and process the annotated documents. The system contains several processing tools like XML elements and attributes addition, deletion, and substitution. For navigation over XML documents the system exploit XPath language. Two main tools of the system are (1) Regular Cascaded Grammars; and (2) Constraints over XML Documents.

Regular Grammars in CLaRK System. The regular grammars in CLaRK System work over token and element values generated from the content of an XML document and they incorporate their results back in the document as XML markup (called *return markup*) (Simov et al., 2002). The tokens are determined by the corresponding tokenizer. The element values are defined with the help of XPath expressions, which determine the important information for each element. In the grammars, the token and element values are described by token and element descriptions. These descriptions could contain wildcard symbols and variables. The variables are shared among the token descriptions within a regular expression and can be used for the treatment of phenomena like syntactic agreement. The grammars are applied in a cascaded manner. The general idea underlying the cascaded application is that there is a set of regular grammars. The grammars in the set are in a particular order. The input of a given grammar in the set is either the input string, if the grammar is first in the order, or the output string of the previous grammar. The evaluation of the regular expressions that define the rules, can be guided by the user. We allow the following strategies for evaluation: "longest match", "shortest match" and several backtracking strategies.

---

[3]http://www.bultreebank.org/clark/index.html

Constraints over XML Documents. The constraints that we have implemented in the CLaRK System are generally based on the XPath language. We use XPath expressions to determine some data within one or several XML documents and thus we evaluate some predicates over the data. Generally, there are two modes of using a constraint. In the first mode **validation**, the constraint is used for a validity check, similar to the validity check, which is based on a DTD or an XML schema. In the second mode **insertion**, the constraint is used to support the change of the document to satisfy the constraint. The constraints in the CLaRK System are defined in the following way: `(Selector, Condition, Event, Action)`, where the selector defines to which node(s) in the document the constraint is applicable; the condition defines the state of the document when the constraint is applied. The condition is stated as an XPath expression, which is evaluated with respect to each node, selected by the selector. If the XPath expression is evaluated as true, then the constraint is applied; the event defines when this constraint is checked for application. Such events can be: selection of a menu item, pressing of a key shortcut, an editing command; the action defines the way of the actual constraint application.

The combination of XLM editor with processing tools is a very powerful tool for minimization of human intervention during the annotation of new corpora. The manual work is inevitable, but many of the mistakes of the automatic processing and also the new annotations are regular. Thus, very quickly the annotator recognizes them. In these cases the system provides necessary support for the annotator to write procedures for automatic repairing or automatic annotation of these regular cases.

At the end a human annotator checks the results and finalizes the annotation. The new information (besides the corrected one) comprises:

- *phrase node (ph)* — subdivision of the sentence with more than one token - bronchial asthma or spine (гръбначен стълб in Bulgarian). It has the term attribute.

- *time string (ts)* — subdivision of the sentence with more than one token for dates and time. It has the time attribute.

- *dosage string (ds)* — subdivision of the sen-

```xml
- <s>
    <tok pos="Nc" offset="1" n="59" lm="квадрипареза" len="13" ana="Ncmsi" aa="Ncmsi" unknown="1" sp="y"
      term="dia">квадрипареза</tok>
    <tok pos="punct" offset="0" n="69" len="1" ana="punct" aa="punct" unknown="1" sp="y">-</tok>
    <tok pos="Af" offset="1" n="60" lm="латентен" len="8" ana="Afsi" aa="Afsi" sp="y">латентна</tok>
    <tok pos="R" offset="1" n="61" lm="за" len="2" ana="R" aa="R" sp="y">за</tok>
    <tok pos="A-" offset="1" n="62" lm="горен" len="5" ana="A-pi" aa="A-pi" sp="y">горни</tok>
    <tok pos="Nc" offset="1" n="63" lm="крайник" len="8" ana="Ncmpi" aa="Ncmpi" sp="y" term="org">крайници</tok>
    <tok pos="Cp" offset="1" n="64" lm="и" len="1" ana="Cp" aa="Cp" sp="y">и</tok>
    <tok pos="Vpp" offset="1" n="65" lm="умерен" len="7" ana="Afsi" aa="Afsi;Vpptcv--sfi" sp="y">умерена</tok>
    <tok pos="R" offset="1" n="66" lm="в" len="1" ana="R" aa="R" sp="y">в</tok>
    <tok pos="A-" offset="1" n="67" lm="долен" len="5" ana="A-pi" aa="A-pi" sp="y">долни</tok>
    <tok pos="Nc" offset="1" n="68" lm="крайник" len="8" ana="Ncmpi" aa="Ncmpi" term="org">крайници</tok>
    <tok pos="punct" offset="0" n="69" len="1" ana="punct" aa="punct" unknown="1" sp="y">.</tok>
  </s>
```

Figure 2: Example 1. Annotation of the upper and lower limbs status

```xml
- <s>
  - <ph term="dia">
      <tok pos="Nc" offset="1" n="136" lm="радикулитис" len="11" ana="Amsi" unknown="1" sp="y">радикулитис</tok>
      <tok pos="Nc" offset="1" n="137" lm="лумбосакралис" len="13" ana="Ncmsi" unknown="1" sp="y">лумбосакралис</tok>
    </ph>
    <tok pos="R" offset="1" n="138" lm="в" len="1" ana="R" aa="R" sp="y">в</tok>
    <tok pos="Np" offset="1" n="139" lm="ляво" len="4" ana="Dm" aa="Ansi;Dm" unknown="1">ляво</tok>
    <tok pos="punct" offset="0" n="140" len="1" ana="punct" aa="punct" unknown="1" sp="y">.</tok>
  </s>
```

Figure 3: Example 2. Annotation of medical terms in Latin transliterated in Cyrillic

tence with more than one token for doses –
1+1/2 pill or 125 mcg per day.

Information from the attributes:

- **_term attribute_** — marks the medical terms
  and bears information about their type

- **_term values_** — diagnosis (**DIA**), symptom (**SIM**), status (**STT**), organ (**ORG**),
  body system (**SIS**), medicament (**MED**), test
  (**TST**) and index (**POK**). It is likely for more
  to come up.

- **_time attribute_** — bears information about absolute (**abt** value) time (10.02.1999) or relative (**rtt** value) time (two months ago).

We apply various vocabularies which help us to
figure out the semantics of the words in the near
context.

The 10 vocabularies are: *(1)* Vocabulary of the
100,000 most frequent Bulgarian terms (Osenova and Simov, 2010); *(2)* Generic medical terms
in Bulgarian; *(3)* Anatomical terms in Latin; *(4)*
Generic names of drugs for Diabetes Mellitus
Treatment; *(5)* Laboratory tests; *(6)* Diseases; *(7)*
Treatment; *(8)* Symptoms; *(9).* Abbreviations;
(10) Stop words;. These are applied in the specified order and the annotations of the latter ones
override the previous ones. The vocabulary coverage is shown on Table 1. In the columns are shown
the size of each vocabulary (Size) and the number
of tokens matched in the text by this vocabulary

Table 1: Lexical Profile Statistics.

Category	Size	Tokens
1. btb	102,730	41,582
2. bg med	3,624	1,545
3. term anat	4,382	3,792
4. drugs	154	12
5. lab test	202	18
6. diagnoses	8,444	54,431
7. treatment	339	4,170
8. symptoms	414	4,180
9. abbrev	477	14,404
10. stop words	805	67,153

(Tokens). The largest coverage has the vocabulary
of stop words, then diagnoses, next is the vocabulary of most frequent Bulgarian words followed
by the markup words.

4   Experiments and Results

The experiments were done over a set of 1,370
pseudoanonymised discharge letters in Bulgarian
for patients with Endocrinology and Metabolic
disorders. The discharge letters text contains medical terminology in Latin alphabet (about 1% of
all term tokens in our present corpus), sometimes
with different transcriptions in Cyrillic alphabet.
There are specific term abbreviations both in Bulgarian and Latin (about 3% of the tokens), numerical values (16% of the tokens) and about of 1% of
all term tokens are presented as abbreviations.

One of the main problems is that huge groups of out of the vocabulary terms are available in the discharge letters. They are several groups - medical terms in Latin, medical terms in Latin transliterated in Cyrillic; brand names of drugs and medications, abbreviations, etc. There are 7,108 occurrences of drug names in 1,213 of the discharge letters, in average 5.86 drugs per document. These is a quite dynamic information that needs to be updated monthly and the annotation tool also will lack some information.

The problem of Latin written in Cyrillic is about fast and decent annotation by people without knowledge of medical Latin.

статус пост адреналектомиам билатералис(status post adrenalektomiam bilateralis)
аденомектомиам транссфеноидалем ет телегаматерапиам(adenomektomiam transsfenoidalem ет telegamaterapiam)
статус пост тиреоидектомиам про карцинома папиларе лоби синистри (status post thyreoidektomiam pro carcinoma papillari lobby sinistri)

The method is simple: to take every phrase separately and look for attributes and phrasal base and prescribe `Adj` to attributes and `N` for the base (Fig. 3). There is created a grammar in CLaRK for automated phrase (**ph**).

Another problem is that there are many typos in the documents and a variety of abbreviations for same terms.

5   Conclusion and Further Work

We report work in progress about annotation of clinical narratives in Bulgarian.The role of the grammars (phrasal grammar) in quality of the analysis and time-saving in the annotation process. Phrases do not improve the morphological analysis. Good morphological analysis and lemma recognition improves the phrasal grammar and speeds up the work process. One of the main problems is that we did no have yet several annotations for each document and inter-annotation agreement is not evaluated.

Further work include some preprocessing of the corpus for spelling errors correction both for Latin and Cyrillic that will help in the automatic processing. Another direction for further work is the training of a domain specific tokenizer and POS tagger and improving of the general tokenizer and tagger. Iterative enrichment of the vocabularies after the manual correction of the annotation will also help.

Acknowledgments

The research presented here is partially supported by the grant SpecialIZed Data MIning MethoDs Based on Semantic Attributes (IZIDA), funded by the Bulgarian National Science Fund in 2017–2019. The team acknowledges also the support of Medical University – Sofia.

References

Noha Alnazzawi, Paul Thompson, and Sophia Ananiadou. 2014. Building a semantically annotated corpus for congestive heart and renal failure from clinical records and the literature. In *EACL 2014 Workshop-The Fifth International Workshop on Health Text Mining and Information Analysis, Gothenburg, Sweden, 27 April, 2014, edited by Velupillai, Sumithra and Duneld, Martin and Henriksson, Aron and Kvist, Maria and Skeppstedt, Maria and Dalianis, Hercules.* Association for Computational Linguistics, pages 69–74.

Hercules Dalianis and Sumithra Velupillai. 2010. De-identifying swedish clinical text-refinement of a gold standard and experiments with conditional random fields. *Journal of biomedical semantics* 1(1):6.

Louise Deleger, Qi Li, Todd Lingren, Megan Kaiser, Katalin Molnar, et al. 2012. Building gold standard corpora for medical natural language processing tasks. In *AMIA Annual Symposium Proceedings.* American Medical Informatics Association, volume 2012, page 144.

Jung-wei Fan, Elly W Yang, Min Jiang, Rashmi Prasad, Richard M Loomis, Daniel S Zisook, Josh C Denny, Hua Xu, and Yang Huang. 2013. Syntactic parsing of clinical text: guideline and corpus development with handling ill-formed sentences. *Journal of the American Medical Informatics Association* 20(6):1168–1177.

Rob Koeling, John Carroll, Rosemary Tate, and Amanda Nicholson. 2011. Annotating a corpus of clinical text records for learning to recognize symptoms automatically .

Philip V Ogren, Guergana K Savova, Christopher G Chute, et al. 2007. Constructing evaluation corpora for automated clinical named entity recognition. In *Medinfo 2007: Proceedings of the 12th World Congress on Health (Medical) Informatics; Building Sustainable Health Systems.* IOS Press, page 2325.

Petya Osenova and Kiril Simov. 2010. Using the linguistic knowledge in bultreebank for the selection of the correct parses .

Angus Roberts, Robert Gaizauskas, Mark Hepple, Neil Davis, George Demetriou, Yikun Guo, Jay Subbarao Kola, Ian Roberts, Andrea Setzer, Archana Tapuria, et al. 2007. The clef corpus: semantic annotation of clinical text. In *AMIA Annual Symposium Proceedings*. American Medical Informatics Association, volume 2007, page 625.

Angus Roberts, Robert Gaizauskas, Mark Hepple, George Demetriou, Yikun Guo, Ian Roberts, and Andrea Setzer. 2009. Building a semantically annotated corpus of clinical texts. *Journal of biomedical informatics* 42(5):950–966.

Angus Roberts, Robert Gaizauskas, Mark Hepple, George Demetriou, Yikun Guo, Andrea Setzer, and Ian Roberts. 2008. Semantic annotation of clinical text: The clef corpus. In *Proceedings of the LREC 2008 workshop on building and evaluating resources for biomedical text mining*. pages 19–26.

Aleksandar Savkov, John Carroll, Rob Koeling, and Jackie Cassell. 2016. Annotating patient clinical records with syntactic chunks and named entities: the harvey corpus. *Language resources and evaluation* 50:523.

Aleksandar Savkov, Laska Laskova, Stanislava Kancheva, Petya Osenova, and Kiril Simov. 2012. Linguistic analysis processing line for bulgarian. In Nicoletta Calzolari (Conference Chair), Khalid Choukri, Thierry Declerck, Mehmet Uğur Doğan, Bente Maegaard, Joseph Mariani, Asuncion Moreno, Jan Odijk, and Stelios Piperidis, editors, *Proceedings of the Eight International Conference on Language Resources and Evaluation (LREC'12)*. European Language Resources Association (ELRA), Istanbul, Turkey.

Guergana and Savova. 2017. *Annotating the clinical text - MiPACQ, ShARe, SHAPPn and THYME Corpora*.

Kiril Simov, Milen Kouylekov, and Alexander Simov. 2002. Cascaded regular grammars over xml documents. In *Proceedings of the 2Nd Workshop on NLP and XML - Volume 17*. Association for Computational Linguistics, Stroudsburg, PA, USA, NLPXML '02, pages 1–8. https://doi.org/10.3115/1118808.1118820.

Kiril Simov, Petya Osenova, and Milena Slavcheva. 2004. Btb-tr03: Bultreebank morphosyntactic tagset. Technical report, BulTreeBank Project Technical Report.

Kiril Simov, Zdravko Peev, Milen Kouylekov, Alexander Simov, Marin Dimitrov, and Atanas Kiryakov. 2001. Clark-an xml-based system for corpora development. In *Proc. of the Corpus Linguistics 2001 Conference*. pages 558–560.